The complete entirety of this 8 × 11¼ inch book is dedicated to our ol' buddy Jim Varney, boy genius, good neighbor, and a great bunch of guys.

Jim Varney, BEFORE reading the Book of Knowledge

Jim Varney, AFTER reading the Book of Knowledge

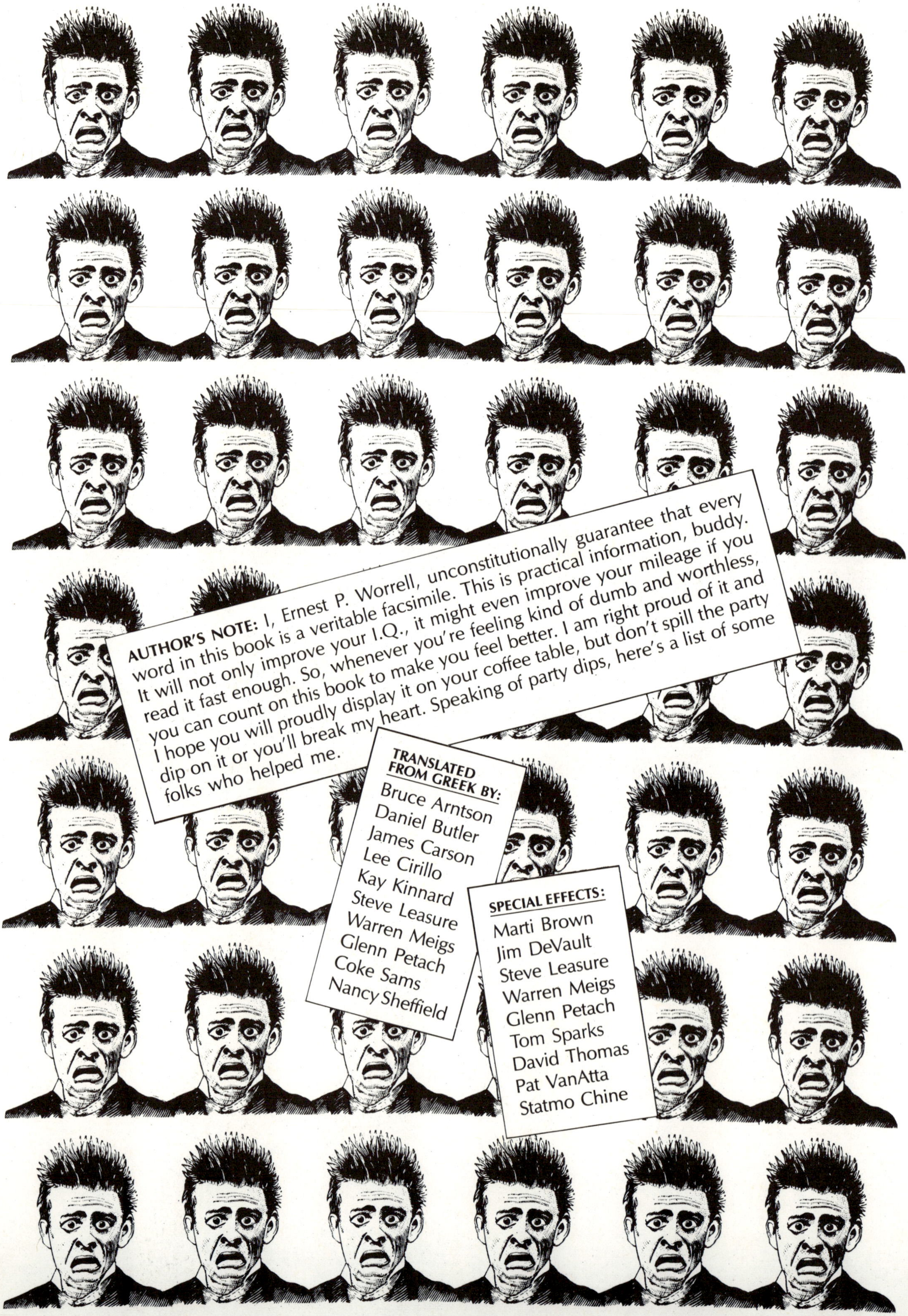

AUTHOR'S NOTE: I, Ernest P. Worrell, unconstitutionally guarantee that every word in this book is a veritable facsimile. This is practical information, buddy. It will not only improve your I.Q., it might even improve your mileage if you read it fast enough. So, whenever you're feeling kind of dumb and worthless, you can count on this book to make you feel better. I am right proud of it and I hope you will proudly display it on your coffee table, but don't spill the party dip on it or you'll break my heart. Speaking of party dips, here's a list of some folks who helped me.

TRANSLATED FROM GREEK BY:

Bruce Arntson
Daniel Butler
James Carson
Lee Cirillo
Kay Kinnard
Steve Leasure
Warren Meigs
Glenn Petach
Coke Sams
Nancy Sheffield

SPECIAL EFFECTS:

Marti Brown
Jim DeVault
Steve Leasure
Warren Meigs
Glenn Petach
Tom Sparks
David Thomas
Pat VanAtta
Statmo Chine

FAVORITE THINGS

- THE RECLINER
- THE FOOTREST
- REMOTE CONTROL FOR TV
- THE TEE SHIRT
- THE BLUE JEAN
- EL BALL DE BOWLING
- CHILI—HOT SAUCE
- MEAT AND THREE
- TAPE DECKS
- FOUR WHEEL DRIVE
- FAST CARS AND FUZZ BUSTERS
- POCKET KNIVES
- OVER / UNDER SHOTGUN
- WORK BOOTS
- BASS BOATS
- POWER TOOLS ←
- PROFESSIONAL WRESTLING
- ROLLER DERBY
- RODEOS
- PRO FOOTBALL
- COLLEGE BALL
- SOFTBALL
- INDOOR SHUFFLEBOARD
- POOL
- PINBALL
- DARTS
- COLOR TELEVISION

Training For Quickness

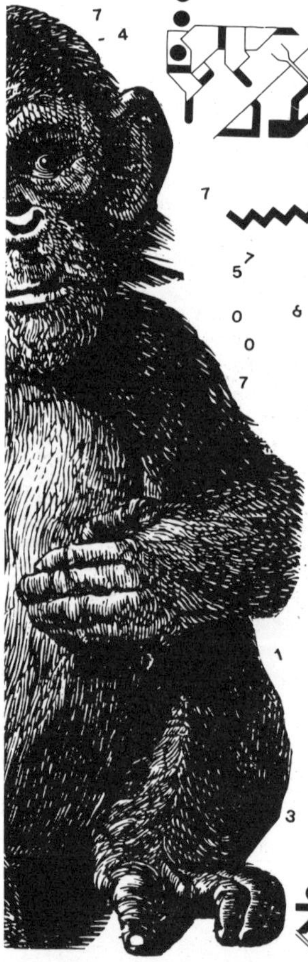

Cousin Lyman taught himself to have real quick hands. You remember that Kung Fu guy on TV and how he would try to grab three pebbles out of the open hand of a blind priest? Lyman couldn't afford a trip to China and didn't know any blind priests, so he practiced grabbing coins away from the blind organ grinder's monkey. The money was good but eventually he realized he was doing wrong and quit. It was right after the monkey picked him out of a police line-up.

FICTION & FACT

- Captain Hook never really lost his hand to the crocodile. He just pulled his sleeves way down.
- Dachshunds are really small crocodiles with fur.
- Cooties did not originate in China, but Czechoslovakia. They have disguised themselves in recent years as round yellow smiling faces that say "Have a nice day." When someone says "Have a nice day" to you, look out! They are really trying to give you cooties.
- There really is such a thing as Krypton and you wouldn't want to mix it into your potato salad.
- The Pygmies invented shuffleboard because they were too short to hula hoop successfully.
- Just as pyrite is the name for fool's gold, Pyrex is the name for fool's glass.
- You, too, can put a fancy degree with your name. For instance, I once knew a man who had "Dr. ________" and "A.B.D.D." on his business card. He said the "A.B.D.D." stood for "All But Doctoral Dissertation." I told him I was Ernest P. Worrell, and that the P. stood for Power Tools. He said I should be Ernest P. T. Worrell but I said that sounded too much like a circus.
- In ancient times, Gaul was a very spicy food eaten only by people who were very nervy, and only in big quantities by big people who were very very nervy. This is why you now hear the expression, "She's got a lot of Gaul."
- The Big Dipper was named for the soup ladle used in the Big House.
- If you do something really awful, you could get sent to the Black Hills of Calcutta.
- Horatio Alger's mother wanted a girl.
- Peeping Tom was really Weeping Tom. He never really saw Lady Godiva—just her horse.
- The first Tinkerbell didn't wear any clothes. The second Tinkerbell wore cobwebs and leaves. The third Tinkerbell had everything dry-cleaned.
- Boarding school is an excellent place to learn how to build things with lumber.
- Magic was first used by the Polish twins Hocus and Pocus, who used to fool people into thinking they were only one person. Later the twins also invented a delightful dance called the hokey-pokey.

FAMOUS PEOPLE WHO ARE FREQUENTLY OVERLOOKED

Hank Cartwright
An American who invented the handle for the water faucet, like in a bath tub or sink. You can see his initials "H" and "C" on a lot of them. Before him, you had to turn the water on and off with a wrench and bang your knuckles when the wrench slipped.

Phillip del Fuego
A Spanish explorer who discovered picante sauce and was honored by Isabella. He returned triumphant to his home town of Pizzar, was kicked in the head by a lame horse, and died. His last words were: "Por que de la gusto cabana" or something.

Pierre LaDeedah
After jumping bail in New Orleans, this 18th century artist hid out in Texas where he created his masterpiece—a free-form design which was eventually used to plot out the current boundaries of that state. At the time, however, they hung him for blasphemy because everyone knew that Texas had no boundaries.

Gustav Asterisk*
The world's leading author of unpublished books, he wrote 1,293 novels and not a one was ever printed. The problem was that he wrote so fast, he kept leaving words out. When his editors complained, he'd write up a list of the words he left out, and then put a little star where they should go. This really made the printers mad, so they got back at him by printing only the little stars and throwing the rest away. Asterisk died of writer's cramp but his work caught on and is still very popular today, especially on coupons and junk mail offering free products.

The Most Dangerous Things In The World

(Memorize the list and then eat this page.)

- Setting your milk glass on the edge of the table.
- Sleeping with gum in your mouth.
- Slicing tomatoes with the blade turned towards you.
- Leaving your shoelaces untied.
- Balancing on the back legs of your chair.
- Wearing dirty torn underwear when there's any chance that you'll be in an accident.
- Going barefoot—if you're wearing dirty torn underwear.
- Reading in the dark.
- Putting something in your mouth if you don't know where it's been.
- Running with anything in your mouth.
- Cutting your toenails too short.
- Not using the handrail, if you're going barefoot.
- Imitating the chimpanzee at the zoo, especially if it has something in its mouth.
- Standing on soap in the bath tub, if your toenails are cut too short.

HOUSEHOLD TIPS

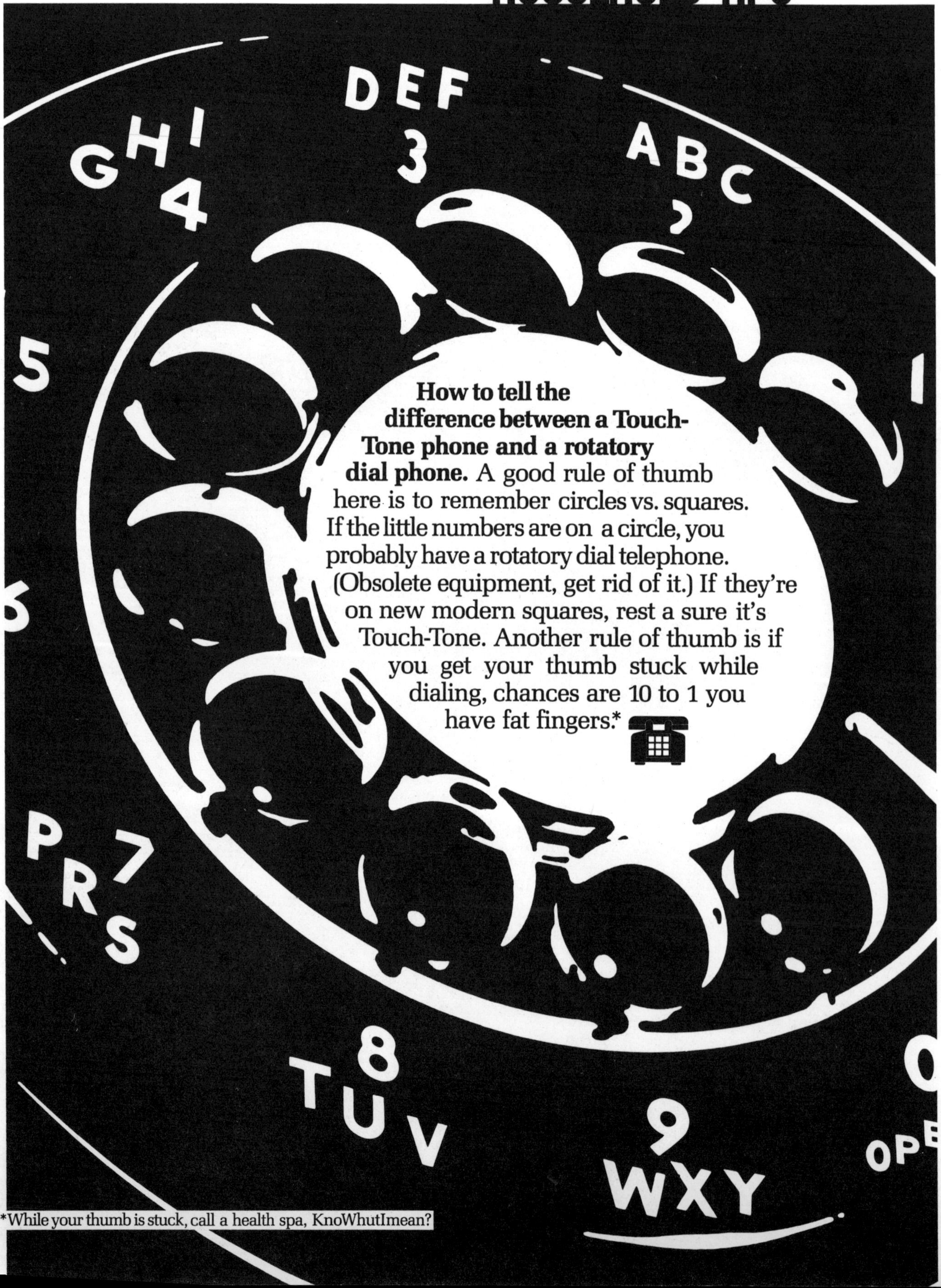

How to tell the difference between a Touch-Tone phone and a rotatory dial phone. A good rule of thumb here is to remember circles vs. squares. If the little numbers are on a circle, you probably have a rotatory dial telephone. (Obsolete equipment, get rid of it.) If they're on new modern squares, rest a sure it's Touch-Tone. Another rule of thumb is if you get your thumb stuck while dialing, chances are 10 to 1 you have fat fingers.*

*While your thumb is stuck, call a health spa, KnoWhutImean?

→ JOB PRESSURE

People ask me, Mr. Worrell, what is job pressure? Well, I recall a story about my old buddy Hurkey Rozette. He and I went to different high schools together. Hurkey has a good job stamping the little M's on M&M candies. You know, them suckers come by at a thousand a minute, and if Hurkey uses too much pressure when he stamps them, it cracks the shell. And even though they're small cracks, Hurkey is in big trouble. This is too much job pressure.

How should you handle too much job pressure? If you work in a candy plant, eat the evidence. If you work in a fertilizer plant, quit. And get a job in a candy plant, KnoWhutImean?

History
PAST AND PRESENT

■ The town of Rugby, North Dakota, is the geographical center of North America and as a result there's not one person there who is presently in analysis.

■ Both Bandon, Oregon, and Gloucester, Mass., claim to be the cranberry capital of the United States and no one from either town has spoken to each other in 47 years.

■ Ed Ames has a twin sister that can't carry a tune.

■ The Osmond Family once challenged the entire Ted Rusker choir to a foot race. The proceeds were to go to charity.

■ Ted Rusker, of Little Rock, Arkansas, put the "wam" in the wam-a-lam-a-ding-dong.

■ In Iran, you can actually be blinded for judging a book by its cover. It is also said of Iran that it is situated between Iraq and a hard place.

■ The Mississippi river is actually the Missouri river, and vice versa, but once the mistake was discovered there were already several maps in circulation.

■ Helen Reddy never wore a dress until she was 19.

■ You could fill an SST with fish from Lake Erie and fly it halfway across the Atlantic before anyone would even notice they were missing.

■ If you put end-to-end all the link sausage purchased in the city of Danville, Illinois (hometown of Dick Van Dyke) over a period of two years, they would reach the neighboring town of Hoopeston and come part way back.

CONTINUED

■ Davy Crockett played in the house band at the Alamo for at least a month before they were attacked.

History
CONTINUED

■ In the middle ages, the "willies" was as dreaded as the plague. It is thought that the first case of the willies in Europe was contracted from sailors working for Marco Polo on his expedition to the Orient. In fact, Marco Polo himself gave a lot of people the willies.

■ Around the bayous of Monroe, Louisiana, the roads become so thick with frogs that the ban on studded tires has been lifted for the last two weeks of June.

■ Albert Cornwall, third Earl of Canterbury, was the first person to note that sometimes there just aren't enough days in the week.

■ Captain John Smith and Pocahontas were never really lovers. In fact, they weren't even going steady. They only dated off and on for a few months.

■ All that glitters was once gold.

■ Thomas Edison actually invented the phonograph in 1874, but never bothered to tell anyone for two years because there just weren't any good records out at that time.

■ More people believe in Santa Claus than in Uncle Sam.

■ Only about five red letter days are actually celebrated each year.

■ Washing your car has nothing to do with whether or not it rains, but you can put chains on your tires and 9 times out of 10 it won't snow.

■ There are parts of the south where it is still considered acceptable to look a gift horse in the mouth.

■ Polka music is still withstood in small pockets of Eastern Europe and Minnesota.

■ The game of basketball was invented in 1891 by Dr. James A. Naismith. He got the idea from watching American Indians of the west play a game called "wompo" in which a small (but fully-grown) sheep is passed about by young braves and thrown through a hole in a large woven basket. Nearby, young squaws would chant the names of their favorite players. Bccause of the obvious advantage of height in the game, the tallest braves were heavily recruited by the tribes who put the most value on the game. So much so, in fact, that academic skills were often overlooked.

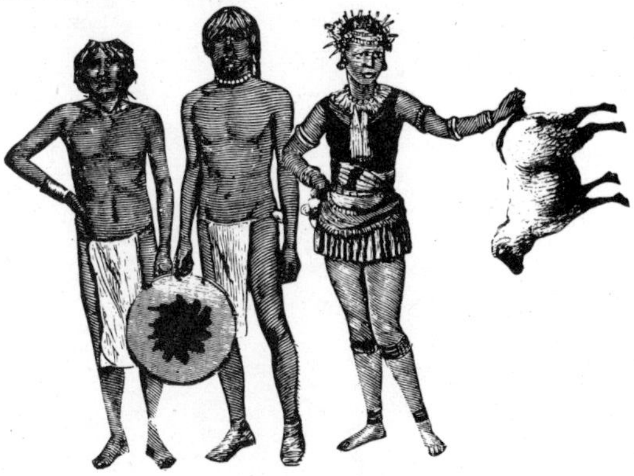

■ The first U.S. flag was made by Ross Ross, but he always insisted his wife Betsy made it, because he was too embarrassed to admit he sewed.

■ Kubla Kahn is known today as a savage conqueror, but in the early days of his manhood he was a popular entertainer, much in demand. Most of modern ventriloquism is patterned after his popular "Kubla Kahn and Ollie" show, in which he would sing while depriving his dummy of water. For a finale, he would beat his dummy's head against the edge of the stage which never failed to bring down the house. Kubla's illegitimate son, Sammy Kahn, penned many popular hits of the day.

■ If you give a dog a fish bone, he'll choke. But if you give a fish a dog bone, it will simply swim away.

■ Kansas is known as "The Gateway to Nebraska."

■ There are parts of Wisconsin where cheese is so plentiful they can't give it away.

■ In the early days of the Oregon Territory, two-thirds of the towns were named by a trapper-politician named Peter Haskel. Many of the towns Haskel named are still in existence, such as Walla Walla, Seattle Seattle, and Spokane Spokane, though some of the names have been shortened through the years.

WHAT TO DO WHEN YOU SPOT A UFO

1. Make sure it isn't a hoax. Pie pans, flocks of birds, and regular commercial airliners have been mistaken for UFO's. Be sure it clearly says "UFO" on the side or you'll look real stupid later.
2. Don't throw rocks at it. That's no way to act with strangers.

3. Note the exact time of day and the weather conditions. The police will be there soon and they always ask this.
4. Act natural. UFO's can sense fear and will sometimes attack without mercy if they know you're scared. Think about something pleasant, like last year when we all went to Panama City for a week.

5. If captured, give only name, rank, serial number and anything else they want to know. Why pretend? Save yourself.
6. Don't bring them home with you. Sure, they're cute but who knows where those little intergalactic guys have been? Imagine how sick space flu could make you.

which square is bigger?

Spaghetti Maze.

Instructions:
Cook one skinny spaghetti noodle for 25 minutes. Drain well. Put one end of the noodle at "Start" and the other end in Ernest's mouth without touching any lines.

HINT: Don't use tomato sauce.

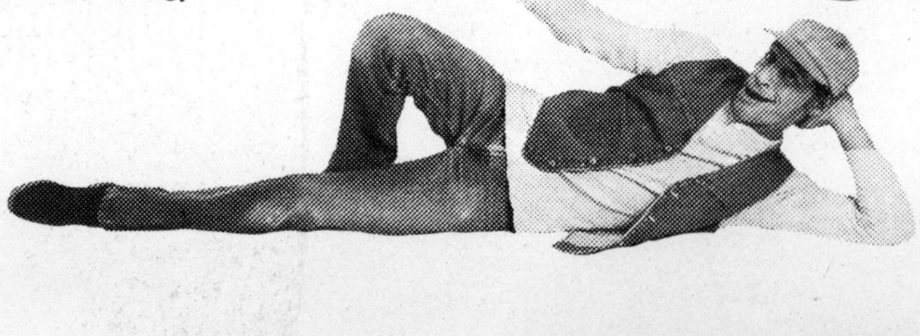

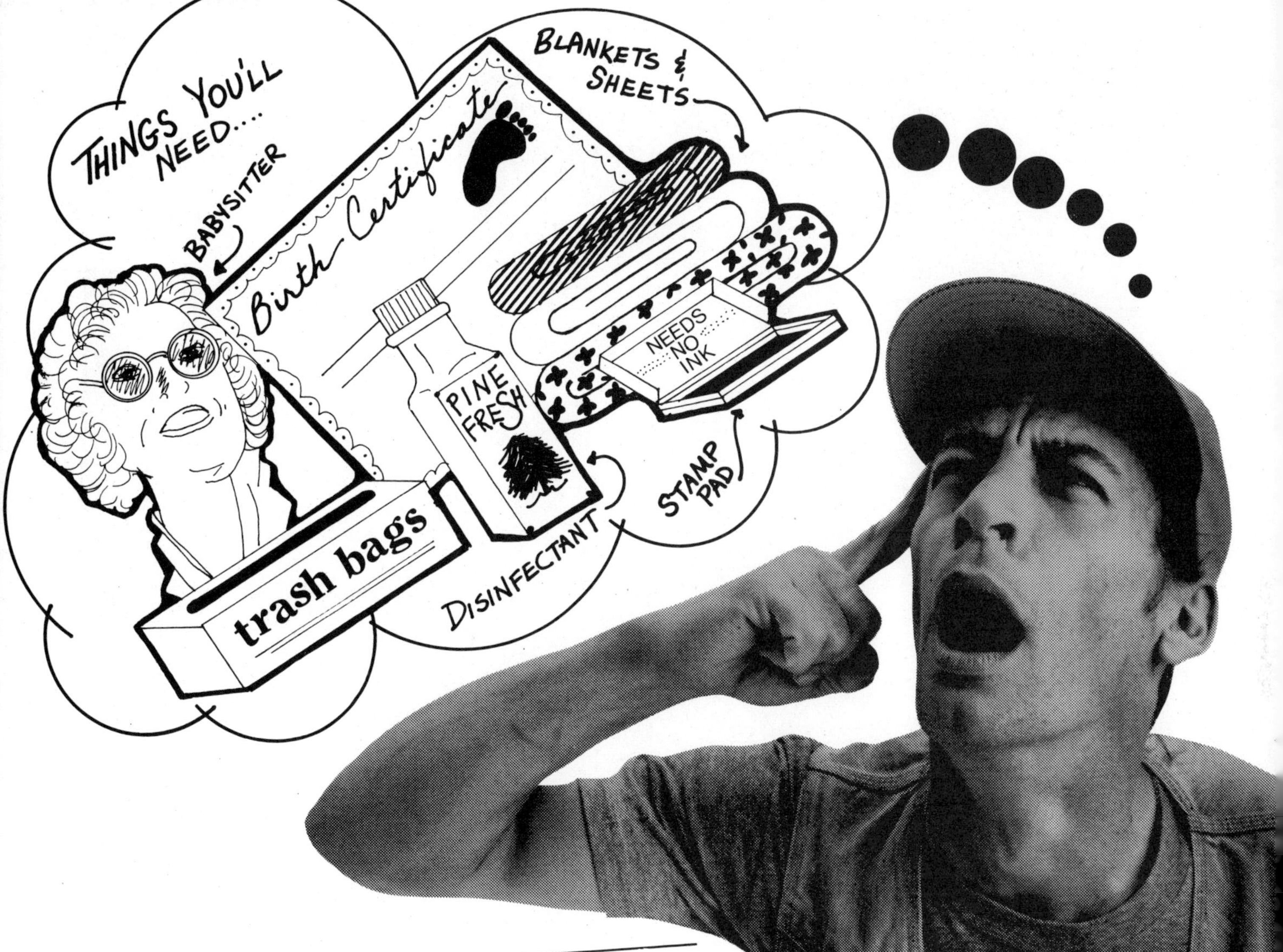

HOW TO DELIVER A BABY ON A STUCK ELEVATOR

This happens on the TV (the real world) all the time, KnoWhutImean? To be prepared for it, your friend and personal adviser, Ernest P. Worrell is going to tell you exactly what to do. (1) Don't get on elevators with pregnant ladies. (2) If you do, be prepared. The equipment you need is a large bottle of pine-scented disinfectant, a blanket, a sheet to tear into strips, a blank birth certificate, an ink pad to take a foot print, plastic trash bags to clean up, and a babysitter.

When the emergency starts, pine disinfect yourself real good all over. Spread out the blanket and tear the sheet into strips. Then, pine disinfect the mother-to-be. The strips of sheet are to tie her because she's not going to like it. Then, you tell her to breathe hard and push over and over until the baby comes. Then, you pine disinfect the baby real good. You can never be sure where it's been. Then, pay the babysitter and get off on your floor. If you have time, ask to speak to the building superintendent and complain about the mess in the elevator.

FUN WAYS TO WEIRD~UP YOUR LITTLE BROTHER OR SISTER

1. Tell them there really is a bogeyman and he lives under their bed.
2. Talk about the extra money you could make if you rented out their rooms.
3. Put a rubber snake in their closet.
4. Put buttermilk on their morning cereal.
5. Don't tell them when there's a holiday from school—let them get up at the regular time and wait outside for the bus.
6. Pack their lunchbox with a can of green peas and of course no can opener.
7. Put different names on their underwear.
8. Point to the top button on their shirt and then flip their nose when they look down.
9. Offer them a bite of your Twinkie and then put the whole thing in your mouth.
10. Unadjust the horizontal hold on the back of the TV.
11. Change the answers on their homework.
12. Loosen the bolts on their Hot Wheels.
13. After they've had an argument with you, make them shake to make up.
14. At bedtime, hide in the closet and make moaning and groaning sounds. Don't come out until they fall asleep or leave.
15. Let Grandma babysit for the weekend.
16. When visiting old relatives, always make them hug the old people and kiss them on the lips.
17. Put tobasco sauce on their peanut butter sandwich when they're not looking.
18. Don't wake them up on Christmas morning.
19. Always take the first lick off the ice cream.
20. When they're at camp, send them your old shoes instead of cookies.
21. Fake a power shortage halfway through Mr. Rogers' Neighborhood.
22. Buy their shoes too big and tell them they will grow into them.
23. Tell them you don't know where their piggy bank is.

NEVER NEVER LAND
Things You Should Never Do
NEVER NEVER LAND
NEVER Put the cat in the microwave.
NEVER Hook your little brother to the milker.
NEVER Buy high, sell low.
NEVER Plant Kudzu by the garage.*
NEVER Leap before you look.
NEVER Spit into the wind.
NEVER Give yourself a hot foot.
NEVER Date a hypnotist.
NEVER Bite your fingernails.
NEVER Pick your nose and eat it.
NEVER Put a penny in the fuse box.
NEVER Eat Mexican food more than 3 days in a row, unless you're in Mexico.
*I can't stress how important it is NOT to plant Kudzu next to the garage. KnoWhutImean?
*I can't stress how important it is NOT to plant Kudzu next to the garage. KnoWhutImean?

TELLING TIME IN THE DARK

The only one I know that can tell time in the dark is my dog Shorty. He gives me the correct time of night using various methods. When he sticks his cold, wet nose under the blankets and licks my face, I know it's time to let him outside. When he slinks up under the bed and whines, I know it's time to let the cat outside. And when he takes a running leap up on the bed and plunks his hind quarters on the pillow, I know it's time to get one of those LSD clocks with numbers that flash on my eyelids all night long and make me dream about going down to the truck stop for some meat and three.

WINNING AT DOMINOES

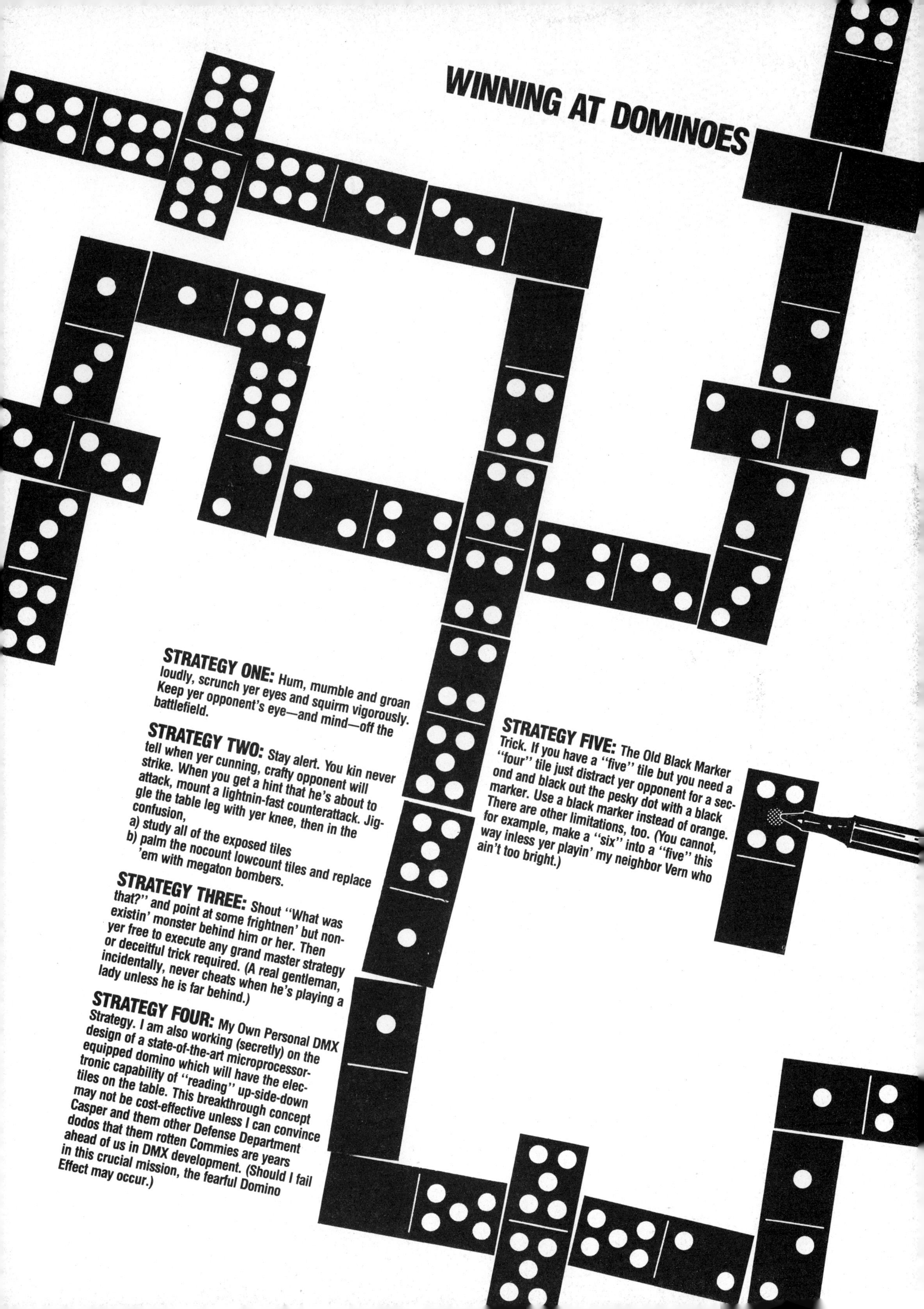

STRATEGY ONE: Hum, mumble and groan loudly, scrunch yer eyes and squirm vigorously. Keep yer opponent's eye—and mind—off the battlefield.

STRATEGY TWO: Stay alert. You kin never tell when yer cunning, crafty opponent will strike. When you get a hint that he's about to attack, mount a lightnin-fast counterattack. Jiggle the table leg with yer knee, then in the confusion,

a) study all of the exposed tiles
b) palm the nocount lowcount tiles and replace 'em with megaton bombers.

STRATEGY THREE: Shout "What was that?" and point at some frightnen' but non-existin' monster behind him or her. Then yer free to execute any grand master strategy or deceitful trick required. (A real gentleman, incidentally, never cheats when he's playing a lady unless he is far behind.)

STRATEGY FOUR: My Own Personal DMX Strategy. I am also working (secretly) on the design of a state-of-the-art microprocessor-equipped domino which will have the electronic capability of "reading" up-side-down tiles on the table. This breakthrough concept may not be cost-effective unless I can convince Casper and them other Defense Department dodos that them rotten Commies are years ahead of us in DMX development. (Should I fail in this crucial mission, the fearful Domino Effect may occur.)

STRATEGY FIVE: The Old Black Marker Trick. If you have a "five" tile but you need a "four" tile just distract yer opponent for a second and black out the pesky dot with a black marker. Use a black marker instead of orange. There are other limitations, too. (You cannot, for example, make a "six" into a "five" this way inless yer playin' my neighbor Vern who ain't too bright.)

...Wearin' Gucci loafers—with or without the new penny.

...Caught listenin' to Punk Rock or Cross-Over (I'm pure country, Vern. The real thing, genuine, Honest to Loretta. It's a matter of principal, Vern.)

...At any fancy, schmanzy opera—unless it's the ohriginel Grand Ole Opry in Music City, U.S.A., cultural center of the Civilized World.

...Drinkin' a Pink Elephant, Golden Sunrise, Rainbow Delite or anything that don't come in a pop-top can or bottle.

...Walkin' into a Unisex barbershop, Unisex clothes store or Unisex anythin. If God had wanted men and women to spend so much time together, why did he invent fishin', checkers or the Miami Dolphins?

...Goin' to any movie with subtitles, cept maybe one of Clint Eastwood's spaghetti westerns.

...In one of them fern bars where ya can't tell the difference between the waiters and the waitresses.

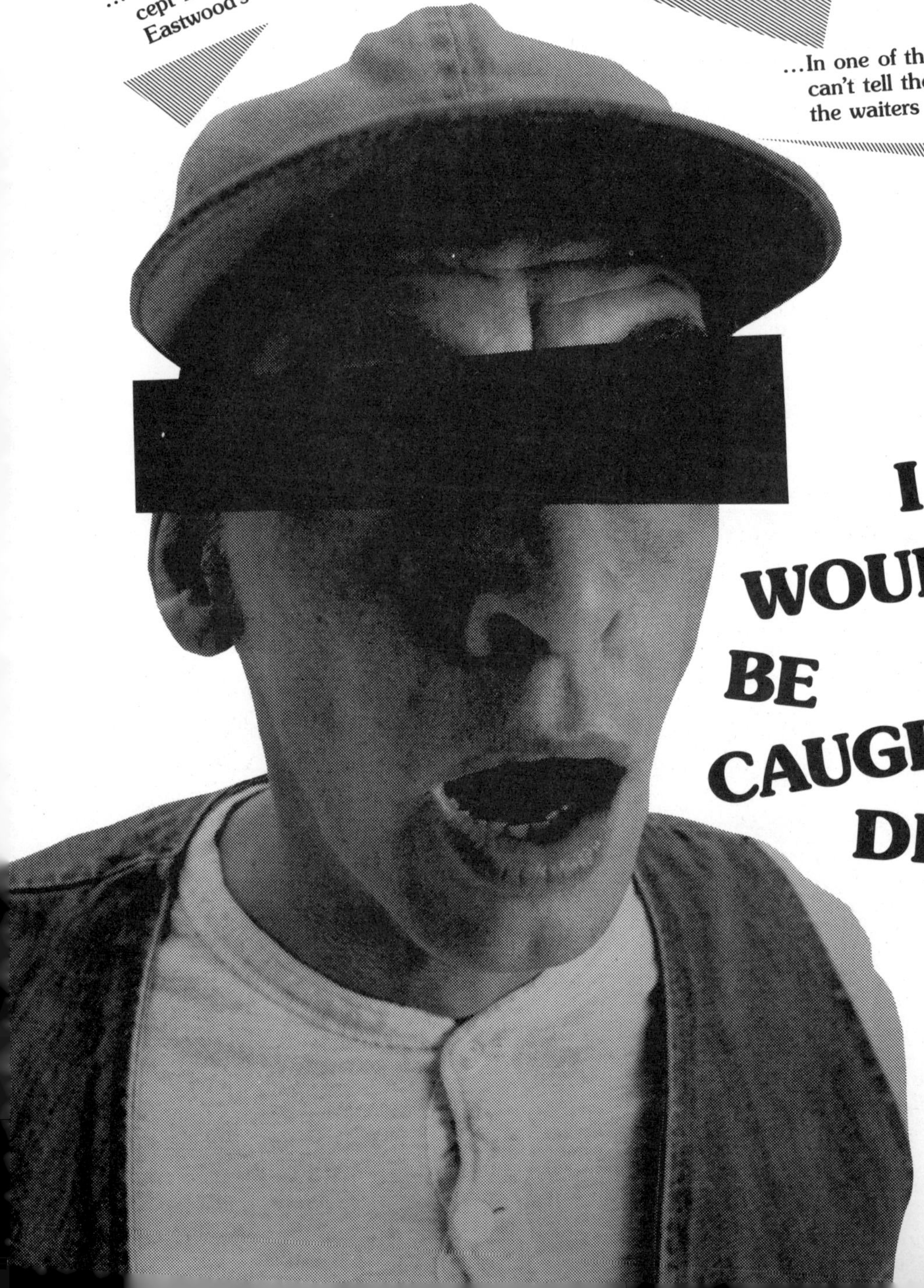

...Playin' tennis, racquet ball, bridge or any of them sissy games.

I WOULDN'T BE CAUGHT DEAD

...Askin' anybody in the entire world for an autograph, cept maybe Waylon or Willie.

...Missin' a single episode of Jane Fonda's exercise program.

...With a spoonful of Queesh, Shrimp Saysjuan or any of that Cuiseen Newvo. Good old burgers, fries or meat and three. KnoWhutImean?

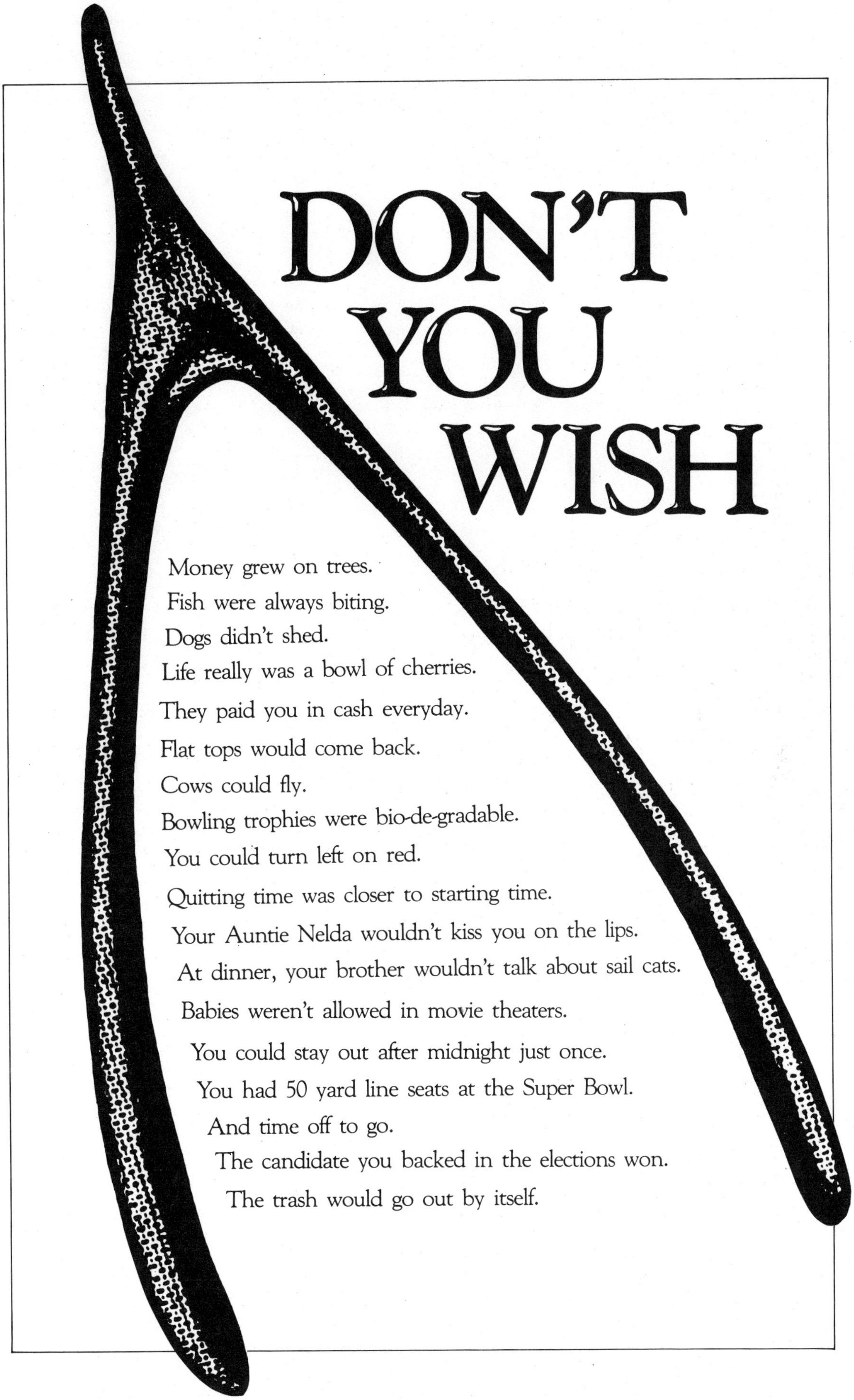

DON'T YOU WISH

Money grew on trees.
Fish were always biting.
Dogs didn't shed.
Life really was a bowl of cherries.
They paid you in cash everyday.
Flat tops would come back.
Cows could fly.
Bowling trophies were bio-de-gradable.
You could turn left on red.
Quitting time was closer to starting time.
Your Auntie Nelda wouldn't kiss you on the lips.
At dinner, your brother wouldn't talk about sail cats.
Babies weren't allowed in movie theaters.
You could stay out after midnight just once.
You had 50 yard line seats at the Super Bowl.
And time off to go.
The candidate you backed in the elections won.
The trash would go out by itself.

Thanks We Owe The Indians

▶ **Do you realize when you are eating popcorn at the movie that we owe thanks to the Indian Chief Kahneewhapoo? In English, his name means Buttery Flavored Cooking Oil. The Chief's technique for making popcorn was to fill a white man's pockets with the kernels and burn him at the stake. Of course, when the movie theatres came along, they changed all that. The Chief had the show and the concession in the same place.**

▶ **When I, myself, order popcorn at the movies, and they ask if I want butter on it, I always say—Butter? On *my* popcorn? Why treat me special? Just give me that yellow grease you're putting on everyone else's.**

Great Questions

GREAT QUESTIONS THAT ARE PONDERED

If Mommas shudn't let their boys grow up to be cowboys, what shud they be?

Has the Tooth Fairy come out of the closet yet?

Does that little bitty light really stay on when you close the refrigerator door?

Kin you tell me why Tennessee is south of Kentucky? And, to really get to the point, why is it east of Arkansas?

Do still waters really run deep?

How come it only snows in the wintertime?

Do you suppose those Indians wud take Manhattan Island, the mugging capital of the world, back for twenty four bucks and some beads?

What ever happened to Jack Armstrong?

What would happen if you snipped off that tag on the mattress?

Why is Vern so dern camera-shy?

Who do Pollacks tell jokes about?

Why does my friendly banker charge me 18% interest on my Charge Card if this is a free enterprise system?

Where do all them friendly bankers go when yer payment is a couple days late?

Can anybody tell me why puppies have to grow up?

Was the person who said that "The way to a man's heart is through his stomach" just near-sighted?

If you are whut you eat, how come ole Ernest doesn't have a pop top? KnoWhutImean?

What are ya supposed to think about a guy who drives around in a car with a pink sign that says "Ask me about Mary Kay Cosmetics"?

HOW TO SPEAK INTERESTING OR USEFUL PHRASES— ENGLISH

1. You're standing on my foot.
2. I was going to park there.
3. More gravy, please.
4. I don't jog.
5. I didn't know that was your dog.
6. Pass the catsup/ketchup.
7. Nobody told me about it.
8. I'll bring it in tomorrow.
9. I didn't know that was your sister.
10. I meant to unplug it. (I thought it was unloaded.)
11. Your tie is in your soup.
12. The check is in the mail.
13. I could have sworn today was Tuesday.
14. The dog ate my homework.
15. The alarm clock didn't go off.
16. Where is the red deck chair?
17. Nice shoes.
18. Can I use your bathroom?

PERFECT MARRIAGE

My friend Bob married a waitress because he thought it would make the perfect marriage, KnoWhutImean? You see, every time she did something for him, he tipped her. He even kept coins in the soap dish in case he had to ask her to bring him a clean towel. It all worked real well for about a year until he noticed the other guys at the restaurant one night, and they were bigger tippers. His wife left with one of them. Bob wanted to stop her, but he was trying to straighten out his check that was mixed up with someone else's.

Bob had ordered the short ribs and collard greens, but he didn't have the corn-on-the-cob or apple pie with a slice of cheese. The other guy claimed he ordered the lemon meringue but didn't get it. Bob never went back to that restaurant again and to this day he refuses to eat short ribs and collard greens.

PEOPLE WHO ARE NAMED FOR FOODS

- PEANUT GILLESPIE, PAPER BOY
- PEACHES SEDBERRY, PROFESSIONAL DANCER
- KRYSTAL GAYLE, FAMOUS SINGER, LORETTA'S SISTER
- CATFISH HUNTER, BASEBALL PLAYER
- KILGORE TROUT, SCIENCE FICTION WRITER
- HAMILTON BURGER, DISTRICT ATTORNEY, "PERRY MASON", LOSER
- TED EGG, RETIRED SHORT ORDER COOK
- BABE RUTH, BALL PLAYER
- FRANCIS BACON, DEAD PHILOSOPHER
- HONEY MALONE, WAITRESS
- SPUD TUCKER, OBSCURE SINGER

What every woman should know about buying batteries?

over →

HOW TO INSULT YOUR NEIGHBORS THE ERNEST WAY
Where you been, buddy?
Are we communicatin' here?
Who knows, Vern? Maybe we're related! KnoWhutImean?
'Bout time you did a little fixin' up, Vern!
You're lookin' tired, Vernon!
Is *this* the water you drink?
Huh! What'd 'at set ya back?
You been lookin' for love on all the wrong stations!
The way you eat all this stuff, Vernon, I'm surprised you ain't belly up! KnoWhutImean?
Yeah, I'm sittin' here warm 'n comfy while you're payin' the big potatoes! KnoWhutImean?
You spend more time fixin' this ol' heap than you do drivin' it!
Is *that* the station you watch?
'Bout time you got home, Vern!
When you gonna wise up, buddy?
Is *that* your dinner?
Is *that* the car you drive?
Is *that* your lunch?
Is *that* your body?
Now it don't take a mental heavyweight to figure that out... KnoWhutImean?

WHAT TO DO IF YOUR CAR GOES OVER 55 MILES PER HOUR.

☐ Mutter "faster, faster" under your breath.

☐ Fake a Rockford power slide.

☐ Ask everybody how much their life insurance is worth and who gets all the money.

☐ Say you're going to be sick.

☐ Ask everybody if they can hear that funny winding noise.

☐ Cry.

☐ Close your right eye and tell everybody you're driving with both eyes closed.

☐ Turn the windshield wipers on—even if it's not raining.

☐ Turn on your fuzz buster.

☐ Start singing "Dead Man's Curve."

☐ Adjust the mirrors constantly.

☐ Turn up the stereo and play Inna-Gadda-Da-Vidda (the long version).

☐ Count out loud with the speedometer and every time the car goes faster, get louder.

☐ Ask everybody if they have their seat belts fastened, if their doors are locked, and if they flossed this morning.

☐ Look for a rest stop.

2606—DRESS 3—RON 155

☐ Drive safely, be remindful of the weather, and for heaven's sake, keep an eye out for each other.

LEGAL MATTERS

My nephew Raymond went on that television small claims court called People's Court. It was the case of the Leaky Pen In The Father's Best Shirt That The Son Wore Without Asking.

The judge asked Raymond why he wore his father's best shirt without asking, and Raymond says that his father wasn't using it because he bowled on Thursday nights. And then the judge says to Fred (the paternal plaintiff), "Your best shirt is a bowling shirt?" And Fred says to the judge, "It looks better than that long, black dress you're wearing."

Got a little advice on legal matters based on the case. Never be guilty of Contempt of People's Court. The TV judge can't have you locked up, so he makes you wear your underwear on the outside of your clothes until you apologize. Ask Fred, you can't miss him, KnoWhutImean?

Words To The Wise.

Getting It All Together.

Or,

Where To Keep Your Parts.

WHY
I DON'T FLY.
1. WHAT IF SOMETHING WENT WRONG WITH THE PLANE, COULD YOU MAKE IT WITHOUT IT?
2. THINGS THAT LIVE THE LONGEST—LIKE ELEPHANTS, GIANT REDWOODS, WHALES AND TORTOISES—RARELY IF EVER FLY.
3. THE FOOD THEY SERVE IS TERRIBLE. WHAT IF THE SAME PEOPLE ARE FIXING THE PLANE ?

DRESS FOR THE OCCASION: OR PROCEED AT YOUR OWN RISK.

BOWLING: It's a definite fox paw to neglect to wear socks.

AIRPORTS: Roller skates.

PIG ROAST: Wear wings, buddy, you in heaven.

TOLL BOOTHS: The correct change.

THE EMERGENCY ROOM: Clean underwear.

A FISH FRY: Your herringbone suit.

COSTUME PARTIES: Say you went as the invisible man.

TORNADO WATCH: Binoculars and your favorite lead shoes.

GOOFY GOLF: A snorkle mask 'cause that moat will foul you up every time.

POWER TOOLS:

THE DO'S, THE DON'TS, AND THE GRAY AREA

THE DO'S

Do ground everyone who comes into your shop.

Do keep tools well oiled. Overnight submersion is recommended.

Do wear safety glasses, safety apron, safety shoes, and a nice little safety carnation in your lapel so you don't look too industrial.

Do treat your tools with respect, they are your friends.

THE DON'TS

Don't use your band saw to do any magic tricks, use your jig saw.

Don't cut off your fingers.

Don't use your electric drill under water.

Don't dance a jig while jig-sawing.

Don't use a power sander on poison ivy no matter how much it itches.

Don't take your power drill to the bank and ask to see the vault.

THE GRAY AREA

Cute first names for your power tools.

Carving the Thanksgiving turkey with a chain saw.

Sleeping with a variable speed drill under your pillow.

Loaning a neighbor any of your tools.

Offering to do discount dental work.

Shorting the cat's tail.

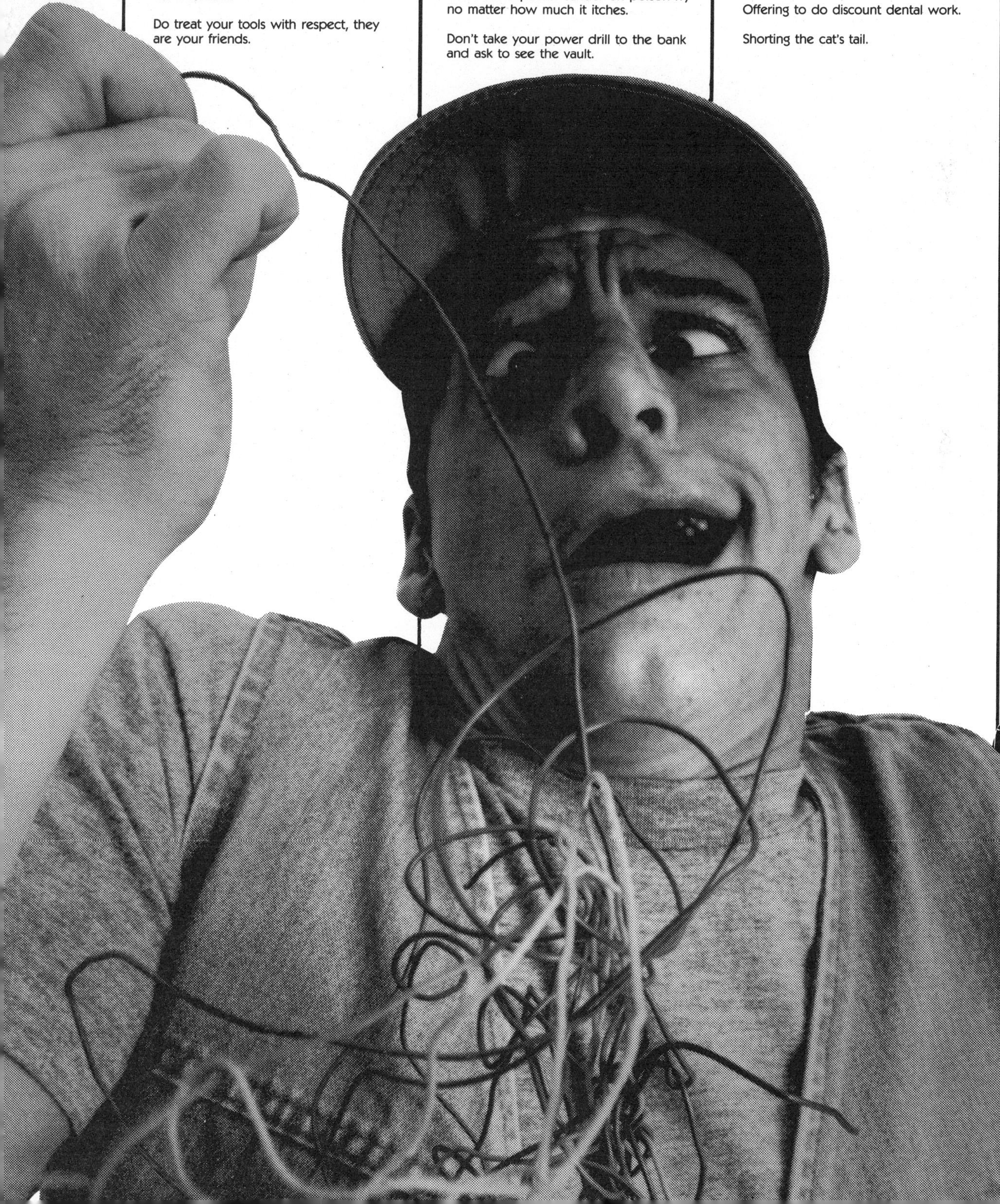

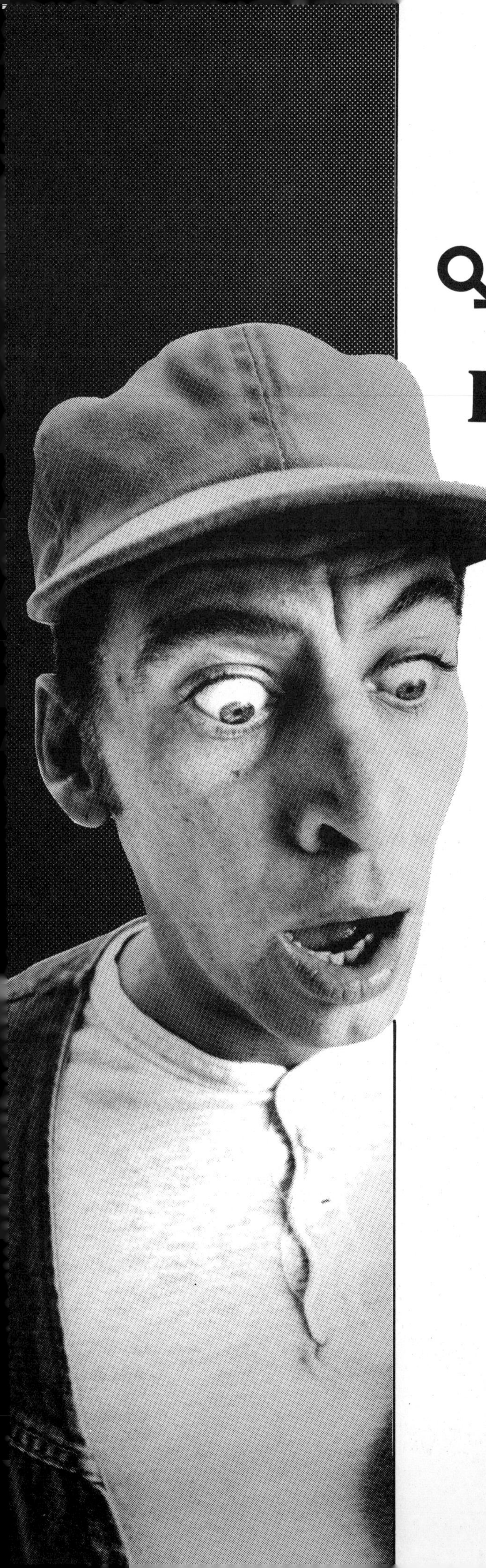

UNLOCKING THE MYSTERY OF PUBLIC RESTROOM IDENTIFICATION?

Saloon keepers and restaurant owners have traditionally added a dimension of fear and terror to the search for public restrooms by using clever titles on the doors instead of crudities like "men" and "women."

As a gesture of good will to other Americans who have never had the opportunity to learn the secret restroom code or a foreign language—through no fault of their own—the Ernest Book of Knowledge includes this handy guide:

MEN	WOMEN
BOYS	GIRLS
MONSIEURS	MADAMES
CABALLEROS	CHIQUITAS
SENORS	SENORITAS
STEERS	HEIFERS
BUOYS	GULLS
US	THEM
STANDING	SITTING

WARNING: A drawing of a human being with long, wavy or curly hair can indicate a Ladies Room...or a hangout for Twinkle Toes and other notorious types.

What's Wrong With This Picture?

THINGS
I REALLY LIKE A LOT

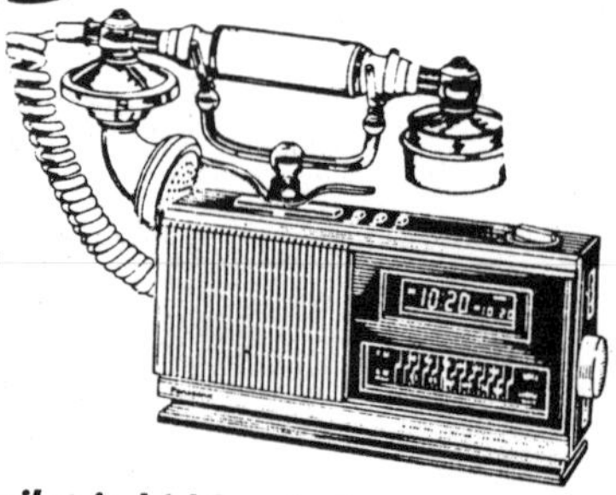

Bathrooms where the toilet is hidden in its own little room behind its own little door.

Pens that are really flashlights.

Clock radios that are really telephones.

Bow and arrows.

Birdhouses that say "See Rock City." And should we not see Mud City?

Rocks that are really places to hide your keys.

Heads of lettuce that are really places to hide your keys.

Electrical outlets that are really places to hide your keys.

Pens that let you write with all different kinds of colors after you press little buttons with the colors on them.

Pencils with my name on them.

Statues of the Statue of Liberty that are really piggy banks.

Gumball machines that are really pencil sharpeners.

Paintings that are really safes.

Secret tunnels.

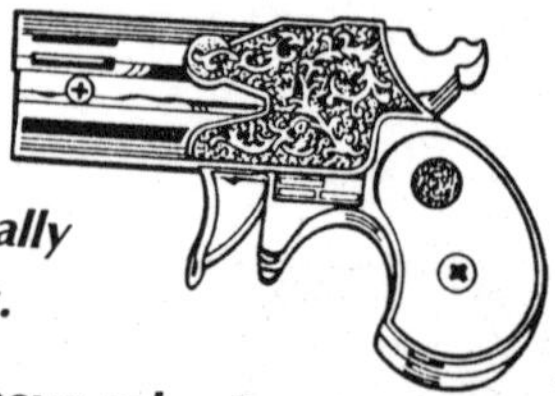

Trapdoors.

Guns that are really cigarette lighters.

Watches that have calendars on the wristband.

That big white building in India—you know, the first shopping mall, built by the Taj. (And when is the guy gonna get credit?)

5
Ways To Frighten Old People

1. Sneak up behind the chair where an old person is sitting. Hook a bicycle chain to one chair leg (or the old person's leg if you're real careful). Take a metal pie pan and run it back and forth over the chain. Watch the old person jump. Big fun.
2. Lie on the sidewalk and pretend you've been run over. (Catsup is optional, but fun.)
3. At dinner, pretend you've got a catfish bone stuck in your throat and are choking to death. Try and turn blue.
4. Bring a lion home from the field trip to the zoo and hide it in the old person's room.
5. Turn the sound down on the TV, then talk to them but just move your lips. They think they've gone deaf.

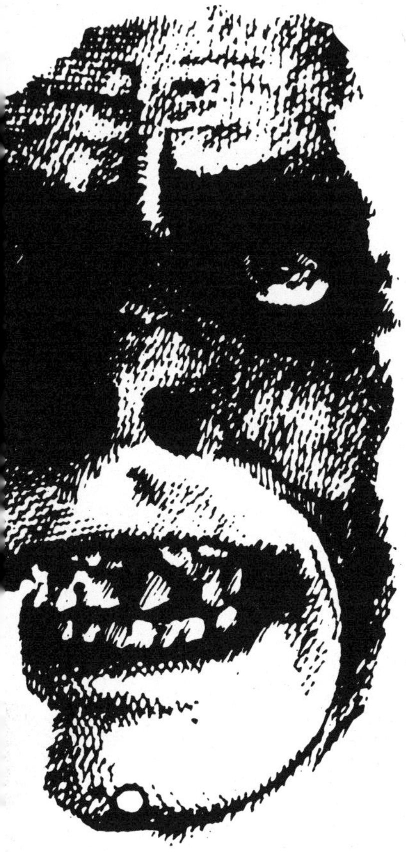

Special Interest Magazines & Coffee Table Decor

Vegematic Age
The latest attachments for your appliance plus many unlikely ways to use it.

Caulking Weekly
Discover 1001 clever ways to put more goop in your life.

Beginner's Catalog of Power Tools
Chain saws with trainer wheels, drills that won't bite, etc.

Carp Quarterly
With many artistic photographs suitable for framing.

Black & White TV Guide
This handy schedule of shows also tells you what color they're wearing—enjoy your black & white set as never before!

Old News Gazette
By the time you read the bad news, it will be over.

Bowling With Cheese
Learn how the pros do it…score bigger…plus (how to photos) show you how to become a gooda bowler.

T-Shirt UpDate
Stay with the trends: French cut is out, never tuck it in.

Bounty Hunters Quarterly
Know where to find the highest priced heads, latest sightings of Nazi War Criminals & vacationing with the wanted. (Free disguise with every issue.)

Poodle Fashions
Adorable outfits for both pet and owner.

Waste Management
Impress your friends with a clear insight of Maggot Wagons (garbage trucks).

Ant Farming
Learn the secrets of life on the hill. Plus raising ants & uncles for profit.

MAKING GOOD THINGS BETTER:

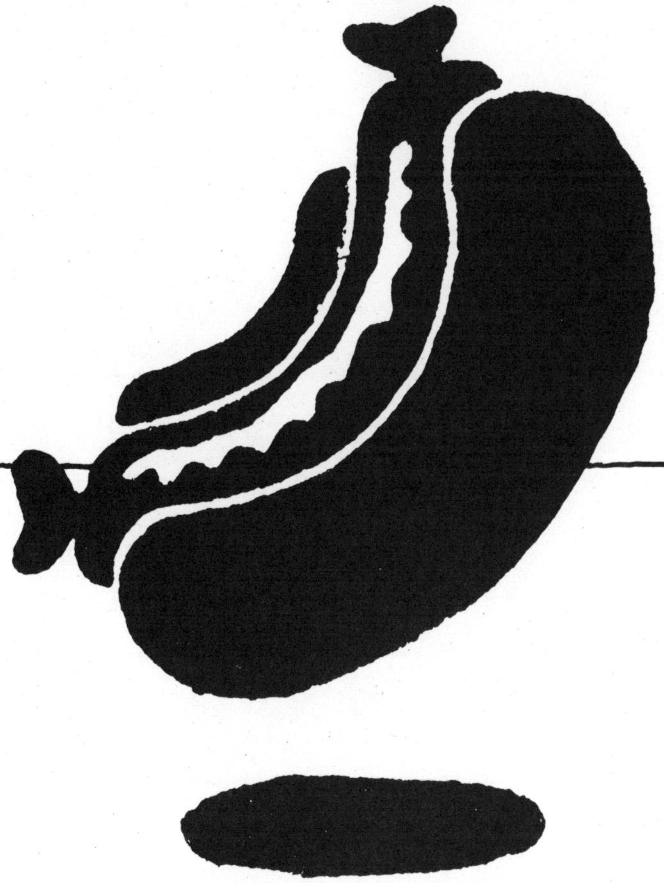

You know as soon as you start boiling your hot dogs, they're gonna start curling up at the ends, right? So when you try to make 'em lie down nice and neat in your standard hot dog bun, they stick out funny and make that first crucial bite sorta awkward and unsatisfying. There's only one way around this nationwide problem. You gotta sew together the ends of your hot dog. Make it into a circle, comprende? Ain't no way that hot dog can curl up then, because it's in the maximum curl mode already. And there's a million other advantages besides. Circular hot dogs can't roll into the crack between your stove and cabinet. Circular hot dogs fit into a little bitty pan without squishing. And you can loop the extras over a Coke bottle for convenient storage. Circular hot dogs are also a perfect fit on bagels and doughnuts and pancakes. There's just no end to the gustatory versatility of the circular hot dog. And all you need is a package of ordinary wienies and a spool of Ernest P. Worrell's Legitimate Digestible Hot Dog Thread. Send for some today and start dining in better circles, KnoWhutImean?

Little Known Turning Points In American History

1706: The arrival in the American colonies of Amos T. Worrell, the legendary progenitor of the famous Worrell clan. (The T stood for Triumph.)

1893: The first industrial use of the circle.

1794: The invention of white bread.

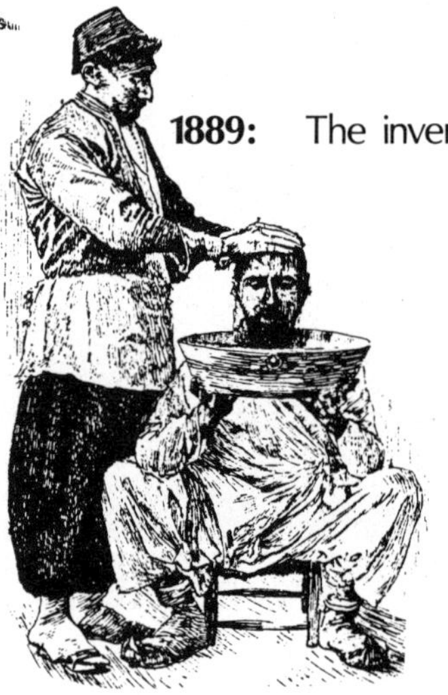

1889: The invention of peanut butter.

1850: First known application of the daily timesheet.

1837: Fashion leader Maudy Worrell designs the first baseball cap—a triumphant style breakthrough that compelled Abner Doubleday to invent the game of baseball two years later.

1973: Self-serve gas stations.

1978: The first genuinely humane application of high technology in modern America: the advent of computer dating.

"TALK BIG AND CARRY A LOTTA CASH."

DOW JONES

Why is the economy in such a mess? It's simple, buddy. It's because the experts keep talking about it. The more they talk, the messier it gets. I think the primo problem is how they talk. Listen to an economical expert for one half minute and your brain will be buzzing with words that nobody else uses. We're talking buzz word bonanza. First off, they tell you about economic indicators. This is a ploy so you won't know they're guessing. Once they have you in their grasp, the fiscal folks lower their voices and start dropping names fast and furious. Like Capital Gains. You get 'em by selling that broken chain saw cluttering up the workshop. Or Deficit Spending, which was once called writing bad checks. I perked up when I heard about Cash flow 'til I found out the only way to get it is to do some Deficit Spending. So you say you need a loan because a rise in credit demands has caused a shortfall in your annual yield. Do not ask for the Preferred Interest Rate. You may prefer it, but you ain't getting it. When they're really stuck, the experts blame the whole mess on the Federal Reserve. Now, buddy, you tell me how the Indians could have caused all this inflation!

SECRETS OF THE PYRAMIDS

1. THEY SNORE AT NIGHT.
2. THEY NEVER RETURN LIBRARY BOOKS.
3. THEY TALK ABOUT YOU BEHIND YOUR BACK.
4. THEY CAN'T MAKE CHANGE.
5. THEIR REAL NAMES ARE EGBERT, LEON AND CHESTER.
6. THEY CAN'T NAME ANY STATE CAPITALS.
7. THEY WERE NEVER IN THE SERVICE.
8. THEIR HIGH SCORE BOWLING IS REALLY 122, NOT 205.
9. THEY ARE OLDER THAN THEY ADMIT.
10. THEY SLEEP WITH A LIGHT ON.
11. WHEN THEY LOOK IN STORE WINDOWS, THEY'RE REALLY ADMIRING THEIR OWN REFLECTIONS.
12. SOMETIMES THEY WEAR SOCKS FROM THE DIRTY CLOTHES BASKET.
13. THEY FAILED THEIR FIRST SCREEN TEST WITH CECIL B. DeMILLE.

HOW TO REMOVE A BUMPER STICKER

So you backed the wrong Bozo in the election, or you've already hugged your kids until they're lumpy. Now you want to get those stickers off the back of the Chevy.

Try this recipe:

Take ½ cup mayonnaise and blend in ¼ cup of hot chili salsa with jalapenos removed. If you don't have the salsa, you can substitute pickle relish and cayenne pepper. Add cayenne to taste. It's enough when your tongue (that movable muscular structure attached to the floor of the mouth) swells and you can't taste anything anymore. Add a dash of salt and enough of any liquid laxative that is chocolate flavored to make it syrupy. Boil until it is a paste and apply it to stickers while hot. Leave on at least one week. Stickers should fall off after that amount of time if dogs don't lick them off sooner.

By the way, the matte finish it leaves on chrome is permanent. This same recipe is also good for antiquing brass and unclogging sewage lines.

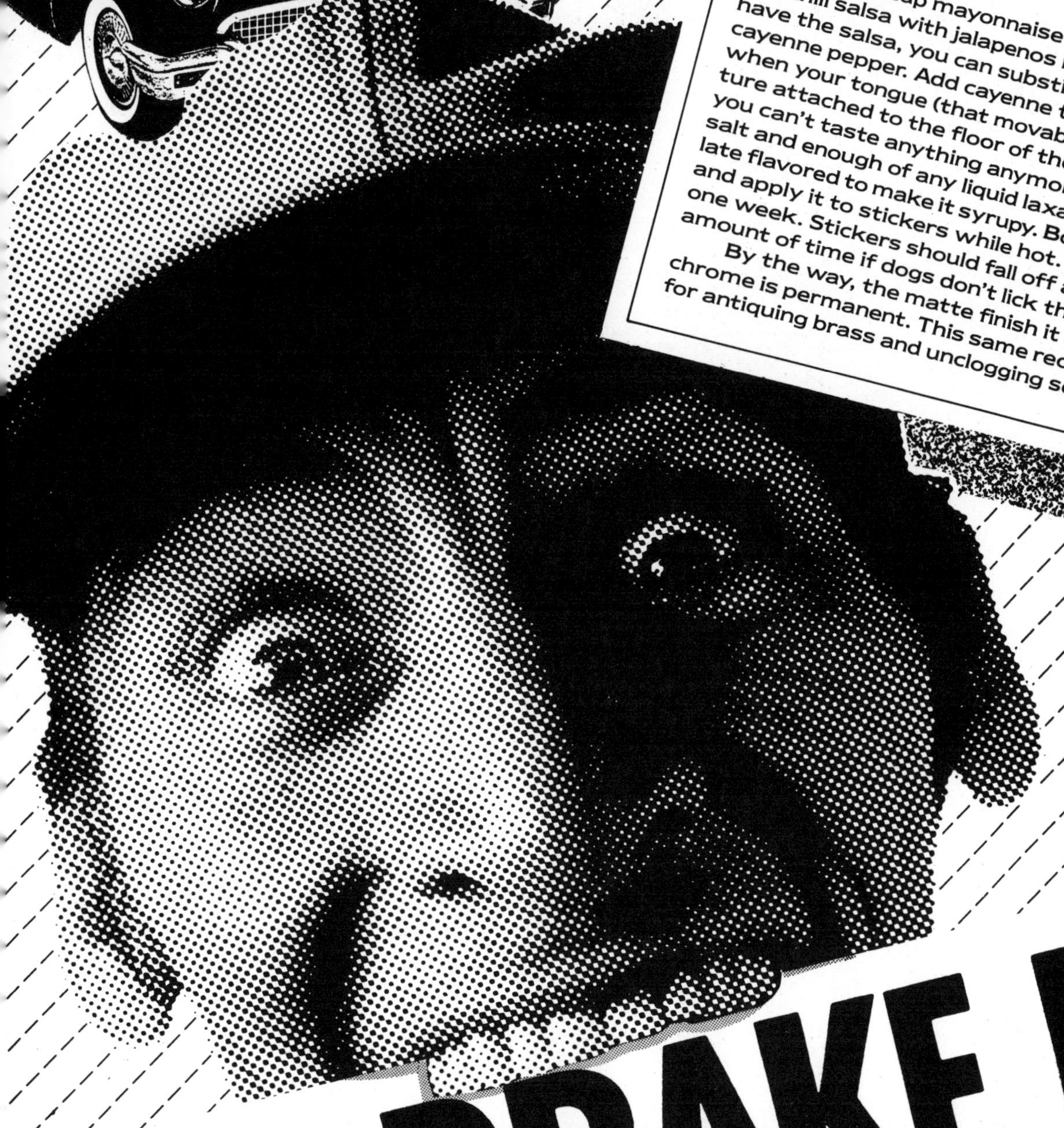

I BRAKE FOR VERN.

KnoWhutImea

HOW TO MAKE BASEBALLS

1. START WITH A LITTLE HARD RUBBER BALL LIKE THE ONE INSIDE OF GOLF BALLS. **2.** GET ABOUT A MILE OF STRING AND WRAP IT AROUND THE LITTLE BALL. (BE CAREFUL, VERN, TO KEEP THE SHAPE ROUND OR YOU'RE JUST ASKING FOR TROUBLE DOWN THE LINE.) **3.** CUT THE LEATHER ARM PATCHES OUT OF YOUR LEAST FAVORITE SWEATER AND MAKE TWO FIGURE-8 SHAPES.
4. WRAP THE BALL/STRING THING IN THE LEATHER AND SEW IT UP WITH RED TWINE. (TWINE IS LIKE STRING ONLY STIFFER AND EASIER TO THREAD THROUGH A NEEDLE.)
5. WITH A BLACK MARKER, WRITE OFFICIAL ERNEST P. "SLUGGER" WORRELL BASEBALL ON THE BALL.
6. KEEP YOUR EYE ON THE BALL.

RIGHT

WRONG

EXCUSES EXCUSES EXCUSES

(FOR JUST ABOUT EVERYTHING)

- The dog ate my homework!
- I was going to do it, right after I finished this.
- The teacher doesn't like me.
- It was raining (snowing).
- I didn't have the change to call.
- But Officer, I didn't see the old lady.
- The alarm didn't go off.
- She didn't mean a thing to me.
- My back hurts.
- Guess I'm just sloppy.
- You didn't show me how.
- I have a headache.
- I couldn't see the blackboard.
- I must have been out of my mind.
- Someone else must have it.
- I couldn't turn left.
- It was meant to be.
- I had to go to Griffin's.
- I forgot.
- It must be in my other purse.
- It must have been my destiny.
- I never heard you say that.
- No one had a watch on.
- I didn't have anything clean to wear.
- I thought you were going to do it.
- I fell asleep.
- The cat ran away and I had to go find it.
- I didn't think it was important.
- All the power at my house was off.
- I couldn't find a pencil.

WASTE NOT WANT NOT

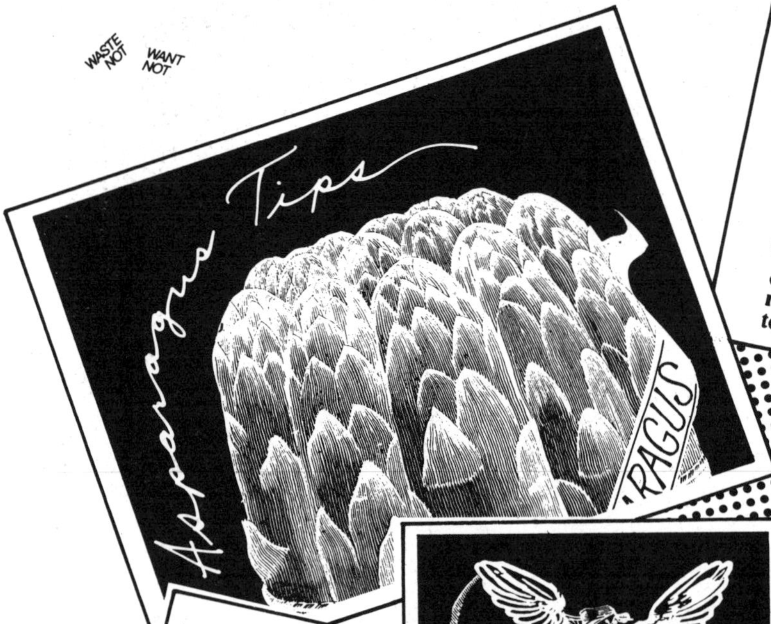

T·I·P·S

tip² (tip) *vt.* **tipped, tip'ping** [akin ? to prec.] **1.** to strike lightly and sharply; tap **2.** to give a small present of money to (a waiter, porter, etc.) for some service **3.** [Colloq.] to give secret information to (often with *off*) **4.** *Baseball a)* to hit (the ball) a glancing blow *b)* to glance off (the bat, glove, etc.): said of the ball —*vi.* to give a tip or tips —*n.* **1.** a light, sharp blow; tap **2.** a piece of secret information [a *tip* on the race] **3.** a suggestion, hint, warning, etc. **4.** a small present of money given to a waiter, porter, etc. for services; gratuity —**tip one's hand** (or **mitt**) [Slang] to reveal one's plans, etc., often without intending to —**tip'per** *n.*

VACATION TIPS

Where to go: Panama City, New Orleans, Nashville.

What to take: Calamine lotion, flip-flops, frisbee, cut-offs.

When to sleep: Do not sleep on the beach your first day out, especially if you wear a church key because it will be sun-tattooed on your chest.

Adapting to change is the secret to survival.

FOREIGN VACATION TIPS

Where to go:

- France—for food cooked with lots of wine and butter (don't eat the snails and say you did), reruns of early Jerry Lewis movies, and buying pretty stuff that doesn't last.
- Italy—for a visit to the Garlic Memorial and buying good shoes.
- Germany—for sausage, beer and a great selection of tools.

Extra Tip: Whatever you do... Don't spell "Tips" backwards. It's not very nice. KnoWhutIMean?

It's the little extras that make the *BIG DIFFERENCE.*

PLACES THAT I FIND IT HARD TO BELIEVE ARE WHERE THEY ARE

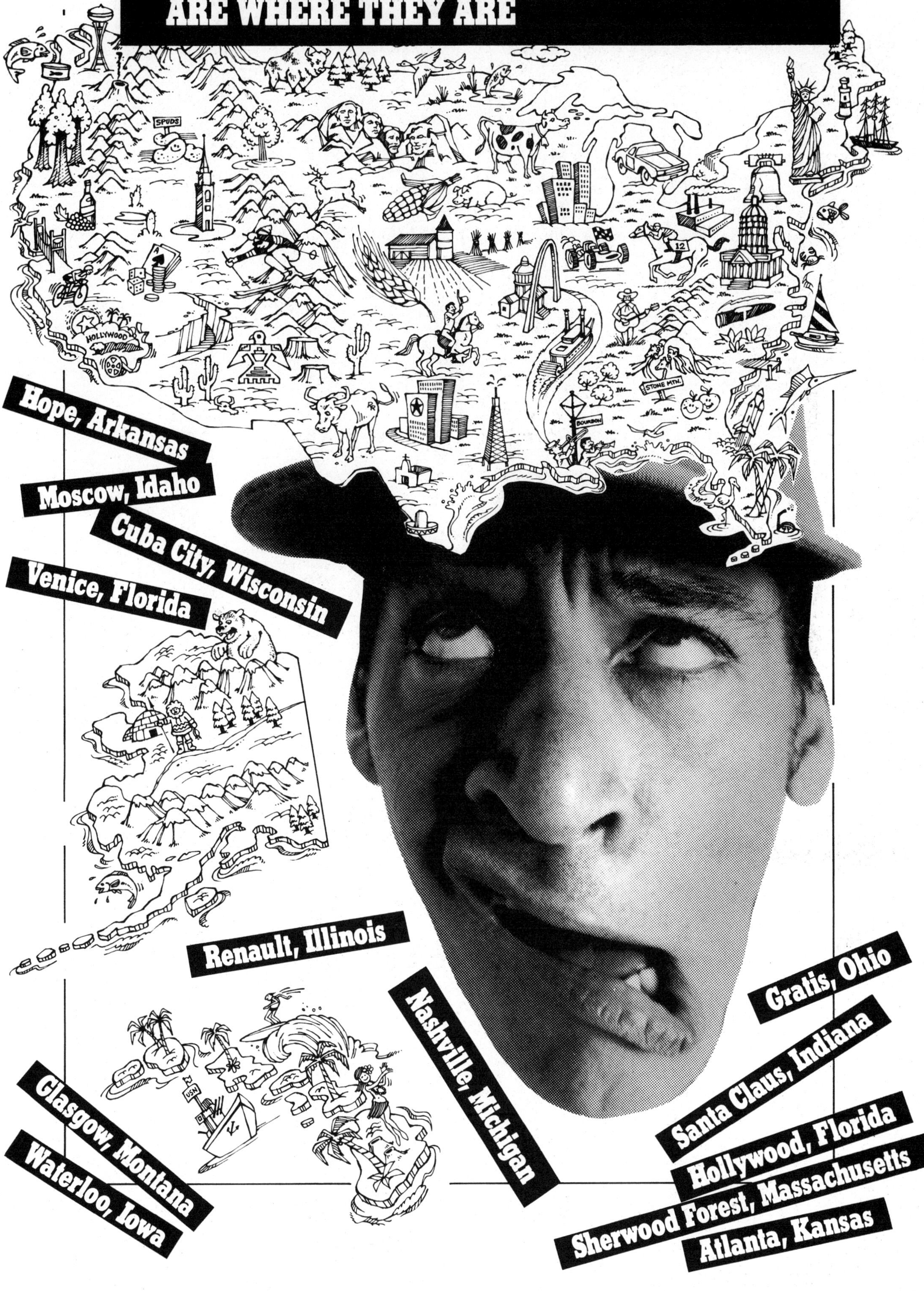

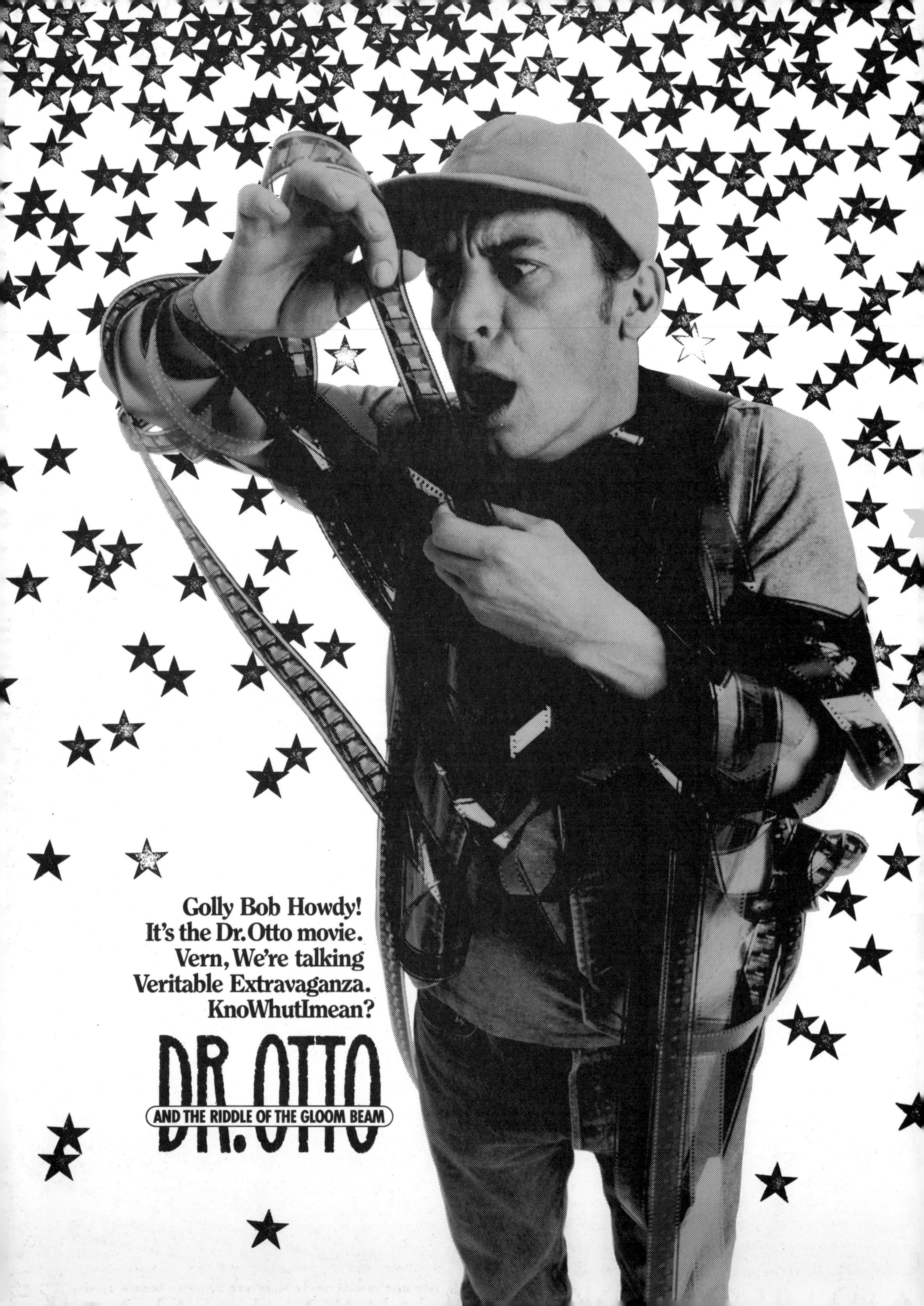
Golly Bob Howdy!
It's the Dr. Otto movie.
Vern, We're talking
Veritable Extravaganza.
KnoWhutImean?
DR. OTTO
AND THE RIDDLE OF THE GLOOM BEAM
DR. OTTO

REAL PLACES WITH NAMES THAT I FIND HARD TO BELIEVE ARE REAL

WAHOO, NEBRASKA
KRUMROY, OHIO
HO-HO-KUS, NEW JERSEY
RONKONKOMA, NEW YORK
KILL DEVIL HILLS, NORTH CAROLINA
OODNADATTA, AUSTRALIA
SEBASTOPOL, LOUISIANA
WHIGVILLE, CONNECTICUT
OUGADOUGOU, UPPER VOLTA
ELMER, NEW JERSEY
PROUTS NECK, MAINE
MAY DAY, KANSAS
TEUTOPOLIS, ILLINOIS
BOGUE CHITTO, LOUISIANA
KOKOMO, INDIANA
HUMANSVILLE, MISSOURI
BUGTUSSLE, TENNESSEE

HOW TO

HOW TO TIE A TIE

STEP ONE:
Open all the long, thin, narrow boxes from under your Christmas tree. Burn boxes, tissue paper and Christmas wrapping in the fireplace.

STEP TWO:
Put the tie over yer shoulders with the fat side on yer left. (See one of these other pages for left/right instructions.) If the tie is one of them tricky knitted ties with no fat end, don't worry which end is which, but try to keep the label on the inside.*

STEP FOUR:
Carefully bring the skinny end up, around and through a loop around yer neck. (This is usually called loopin' the loop.)

STEP FIVE:
Then shinny the knot up towards yer neck until it's snug under yer adam's apple. If you start to feel warm and yer face starts to change colors, you may not understand the definition of the word snug. Ask your wife to loosen the knot while you still can.

STEP SEVEN:
Now go to the bathroom and look in the mirror to make any necessary final adjustments. Ain't you a joy to behold?

**If the label does show, just snip it off—unless, of course, it's a showy Designer Label like Coyote Jim's or Gator City, KnoWhutImean?*

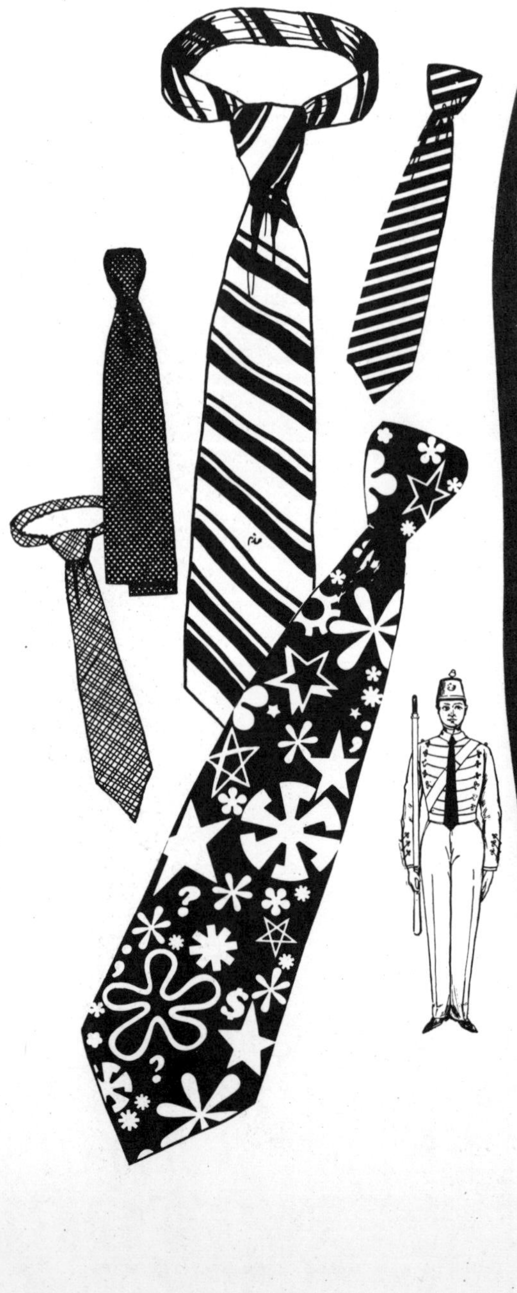

What will the well-dressed neighbor be wearing this year?

BRAIN TEASER

Three men check into a hotel. Desk clerk says: \$10 a piece for the rooms. They pay \$30.00 and go to their rooms. Hotel manager comes by and says: We're having a special, 3 for \$25. Desk clerk gives bellhop \$5 to take upstairs to the men. Going up, the bellhop wonders how to divide \$5 among three guys. He decides to give each guy a dollar bill and keep \$2 for himself. That means each man paid \$9 for his room. (\$10 minus \$1 equals \$9.) But 3 × \$9 = \$27. The bellhop kept \$2 for himself. \$27 + \$2 = \$29. What happened to that last dollar?

The Joy Of Cooking

You are, the Frenchies say, whut you eat. Which explains why them snail-eatin' wogs is so universally disliked worldwide.

But good cookin' is truly the mark of a superior civilization. And yer old buddy Ernest had the good fortune to meet the finest chef in all of America when the U.S. Army sent me down to Fort Pri-mero near San Antonio for basic trainin'. Fat Jack Stubbins was his name and he cooked the finest pot of chili I ever wolfed down.

Fat Jack was a livin' legend in Armed Forces. Grown men begged to be transferred to his battalion because he prepared every plate of food like he was servin' it to the Queen of San Francisco.

I begged Maestro Jack to give me his chili recipe, a secret handed down from his Great Great Grandma Muchas Gracias Garcia Stubbins El Producto III. But he left me hangin' in the intoxicatin' aroma of his legendary chili. Finally, after months of beggin', pleadin' and grovellin' in the Texas dust, I wore the ole coot down and he revealed his family treasure...as long as I promised to maintain the bond of secrecy. But since you and I are ole buddies, I figure that you deserve to share in this mouth-waterin' delight.

RECIPE

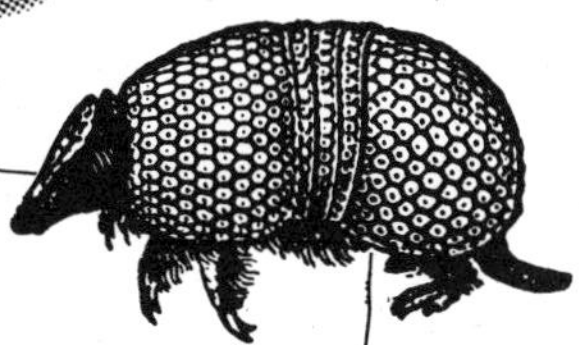

975 lbs. of ground round steak
400 lbs. of pinto beans
36 cloves of garlic, minced
320 medium size onions, chopped
65 lbs. of Jalapeno peppers, quartered
120 tbs. of chile powder
1 medium size armadillo, cleaned
3 lbs. of ground black pepper
1 box of salt
250 cans (12 oz.) tomato sauce
311 ripe tomatoes, chopped
1 package of cheap hot dogs
2 eggs
1 tbs. of corn starch

Brown the ground round steak over medium heat. Add everything else. Cook for a long time.

Serve with crackers and bicarbonate. Feeds 1,500.

Visit the place you first met your loved one and plant kudzu.

Buy a fat, ugly pet and name it after your ex-sweetheart.

Jump off a bridge.

Stare at the sun.

WHAT NOT TO DO IF YOUR HEART IS BROKEN.

Buy a hand buzzer.

Dress up like a clown.

Call them on the phone, then hang up when they answer.

Watch "Casa Blanca" on the late movie.

Listen to sad music with the TV on.

Think about Lover's Leap.

Sulk and mope around.

Join the Peace Corps.

Go shopping at K-Mart on Saturday night.

Hold up curb markets.

Move to Mexico.

Walk with a limp.

Dye your hair.

Burn your clothes.

ANIMAL FACTS

Fact:

THE MEANEST ANIMAL IN THE WORLD

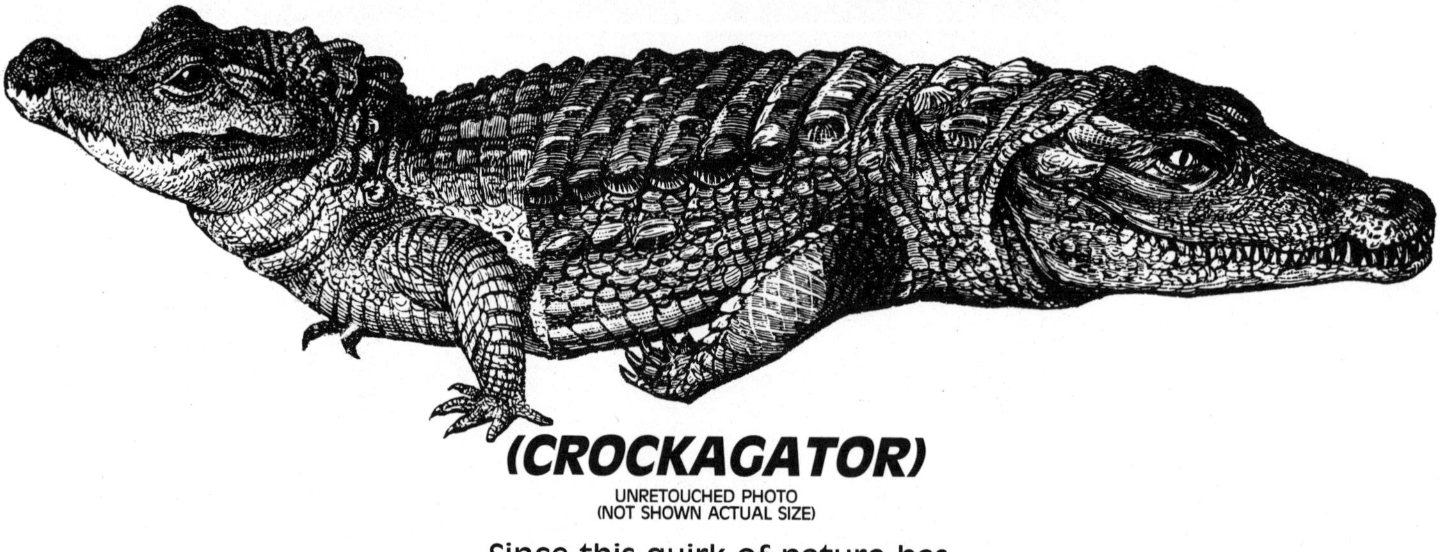

(CROCKAGATOR)

UNRETOUCHED PHOTO
(NOT SHOWN ACTUAL SIZE)

Since this quirk of nature has a crockodile head at one end and an alligator head at the other it's impossible for it to use the bathroom!

(And that's what makes it so mean!)

Fact:

The Suede: is a native of our Great Northwest. These sure-footed deer-like animals have been known to travel hundreds of miles in search of lingonberries which are their favorite food. They are prized for their coats and also for their shoes and handbags. The ones with the least lingonberry stains on them get the biggest prize.

Fact:

The Spam: contrary to popular belief, often does well in captivity and can be trained to feed itself and to enjoy classical music. In the Midwest, herds of spam can still be found roaming the cornfields just as they did when the settlers first found them in the early 1700's.

Fact:

The Wrench: is a direct result of the North American wren and the finch mating while in flight. However the wrench, oddly enough is born without wings and can only consume food while in flight... its life span is a mere 6 hours.

Fact:

Poodles: are of no known value to human beings.

Fact:

Guinea Pig

A guinea pig isn't from Guinea and isn't a pig, but it will still get loose if you leave the cage open, then it will hide in your basement and breed and multiply and be hard to get rid of since you don't want to hurt the fuzzy little guys.

Notes: (write small and take lots of notes...KnoWhutImean?)

SIGNATURE: ______________ DATE: ______________

Ordering From Those Mail Order Catalogs

□ I got one of those mail order catalogs, and it was addressed to Current Resident, KnoWhutImean? It was from Tijuana, Mexico, and it was called *Polyester R Us*. Well, Ernest P. Worrell likes a joke as much as anybody so I sent in an order and signed it "Current Resident." Don't try this at home because it can be hazardous. In fact, in this case two Mexicans named Peppy and Hernando drove all the way up from Tijuana to find Mr. Current Resident. I knew it was trouble when I saw that dusty low-rider car come down the block, and I was really glad I had the foresight not to put my own address on the order.

□ If I hadn't put Vern's address on that order, I could have been the one with all the bruises and abrasions. Those boys were serious. They had three point two million in undelivered C.O.D.'s ordered by Current Resident. Old Vern is in trouble, and I want everyone who could use a polyester poncho to help him out. He'll cut you a real deal on them because as soon as he can sign a check again he has to send Peppy and Hernando a little dinero every month. Pen pals they're not, KnoWhutImean? □

BELIEVE IT OR ~~NOT~~ KNOT

TORNADO BUSTER

Here's a rugged, all-purpose knot you kin use to tie yer chickens down during tornado season. (Be careful to avoid tyin' it around the chicken's neck unless you have an unquenchable cravin' fer chicken frickasea.)

LONG STRONG

This ingenious knot is stronger than Kelvar, Superglue or that stuff them NASA guys use fer nose cones. Tie yer raspberry jello into this here knot and it's as strong as reinforced concrete.

R.I.P.

This complicated knot contributed to the early demise of my Great Uncle Elmo who absent-mindedly put his brand on 33 or 34 of his neighbors calfs.

GUTS

This is the famous knot that folks talk about when they say, "My stomach is all tied up in knots."

GOT ME

Have you ever wondered why anyone on the entire face of the world wud ever need a complicated knot like this? (Me, too. Now whut wud they ever use it fer?)

BOYRDEE

A lot of people use this common, ordinary knot fer tyin' shoes or gift packages, but I prefer to use it to dress up my famous Spaghetti and Meatball Dinners. It's that little extra touch, the imagination, the sheer audacity, that marks the work of a Master Chef, KnoWhutImean?

THE KNIFE

This is the infamous knot that Colonel Nasty used to tie Pearl White to the railroad tracks in them Saturday matinee movies. No one has ever mastered the art of untyin' this incredible knot. (How did them boys ever set Miss Pearl free?)

HANDY GUIDE TO BETTER HABITATION
Ideal Neighborhood

These easy, do-it-yourself improvements can turn your so-so neighborhood into a garden of environmental eptitude. With a couple of pieces of lumber and a staple gun, you can elevate your average block into the realm of Lifestyle.

1 The Hot Dog Stand

Good living starts with good food. If the vendor also can play the accordion, property values could double.

2 Community Land Fill & Car Repair

Dig a hole for everyone's garbage, broken appliances and unwanted mattresses. When it's almost full, it's time for the Neighborhood Oil Changing. Dump the sludge in the hole and light it for a Festive Roast-Your-Own Get-Together. This should be semi-annual.

3 Neon Street Signs

An elegant way to show off your Ideal Neighborhood, especially if it's flashing neon.

4 Uniforms

Your neighborhood will look much tidier when everyone dresses in the Preferred Attire (jeans, T-shirt, and visor cap). Personal color choice is allowed if the local Color Me Beautiful expert is consulted.

5 Indoor/Outdoor Carpeting

Install it from the curbs clear through to the back fences and turn your lawnmowers into go-carts.

6 Bowling Alley

Fashionable neighborhoods are into healthy activities. Choose a straight, flat driveway that's shady.

7 The Tourist Attraction

The Ideal Neighborhood can pay for itself with the right gimmick. Look for ready-made places of interest and hang a sign there:

- Mysterious Cavern (Vern's garage)
- Power Tool Museum (My personal favorite)
- Poodle World (Pen them all up in one place and teach them to be a chorus line)

el·e·phant·skin·daw·g (eléfeňtskińdaw̄g) **n., pl. dawgs,-dawg:** see MUTANT, (Gr. *elephaskindocga*〉(Gen. *elephantoskindocgas),* elephant-skindawg,a medium-sized,thick-skinned mammal, the most medium of extant four-footed animals, with bristle-like hairs, big brown eyes, and foreshortened snout belonging to the same family as the fox, wolf, elephant, jackal, etc.

PLAYING WITH MATCHES

This equation, right or wrong, equals 11. By moving one match, and only one match, make it come out to 130. **Answer is on some other page.*

PLAYING WITH MATCHES — PART 2

Don't be square. This is an easy one. By removing 2 matches, turn this into 2 perfect squares with no extra matches hanging around. **Answer is on some other page.*

For my popcorn, these are the Best Movies Ever.

- ★★★★ 3 Stooges in Outer Space
- ★★★ Gidget Goes Hawaiian
- ★★★½ I Was A Teenage Werewolf
- ★★★ Old Yeller
- ★★★★ Herbie, The Love Bug
- ★★★★★★★ Invasion of the King Crab Legs
- ★★ Beach Blanket Bingo
- ★★★ Viva Las Vegas
- ★★★★½ True Grit
- ★★ Godzilla Meets Bonzo
- ★★½ Mondo Kane
- ½ Short in the Saddle
- ★★★★★★½ Dr. Otto and the Riddle of the Gloom Beam
- ★★★½ For A Few Fistful of Dollars More
- ★★ The Great Houdini
- ★★★ Francis The Talking Mule
- ★★★★ Any Driver Education Film

WHY DO THEY CALL IT........????

1. Why do they call it a runway when airplanes never walk anywhere anyway?
2. Why do they call her Joan of Arc when everyone knows she's from France?
3. Why do they call them fraternal twins when both of them can be girls?
4. Why do they call it cottage cheese when everyone knows it's made in a factory?
5. Why do they call it Virginia when she called herself Elizabeth?
6. Why do they call it fast food but not tell you when you're getting slow food? And are you supposed to eat them at different speeds?
7. Why do they call them goldfish when they're not gold at all—they're orange?
8. Why do they call it The District of Columbia when they could just as well have called it The District of Christopher?
9. Why do they call it rutabaga and is it a cross between a root and a bagel?
10. Why do they call it the Red Sea when it's obvious that it's blue?
11. Why do they call it squash when it's not flat and never will be?
12. Why do they call the part below California "Baja California" but then call the part below Texas "Mexico" and not "Baja Texas"?
13. Why do they call them industrial parks, are they making trees and squirrels and statues inside and if so, why don't they put some outside, of all people they could surely afford it?
14. Why do they call it Standard Candy? And is there an Un-Standard Candy we should be warned about?
15. Why do they call it ham when they could have called it hog?

Little Known Nature Lore

About Bird Brains, Dodos and Twits

The yellow bellied Night Nipper. Its jolly call Tio! Tio! (Uncle! Uncle!) identifies it in the dark, impenetrable Nicaraguan rain forests.

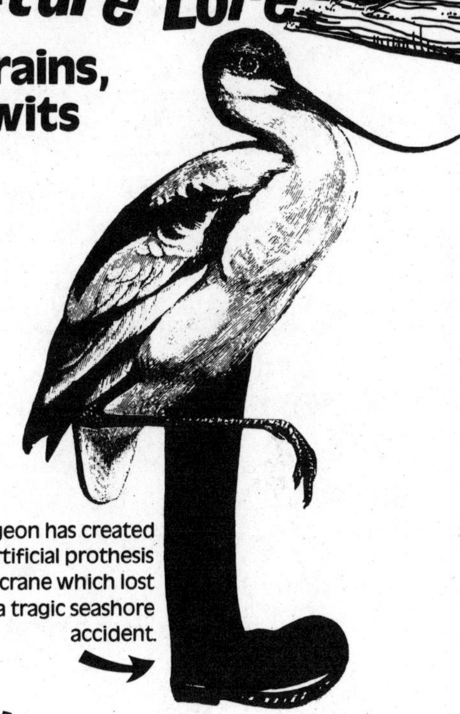

A Boston surgeon has created the first artificial prothesis for a whooping crane which lost its left leg in a tragic seashore accident.

It's a well-known fact that, in the well-organized Bird Kingdom, the male of the species displays the vibrant, colorful plummage. This handsome specimen is a superb example of this unchanging, unwavering law of nature.

This is the thick-headed, red-headed woodpecker which became extinct in the late 1950s when simulated wood-grained aluminum siding became popular in America.

Every decent, law-abidin' American knows that the Swallows return to San Juan Capistrano—regular as clockwork, on March 18th every year—but how come nobody knows that the misunderstood Turkey Buzzard returns to Hinkley, Ohio on April 1st after their winter sabbatical along the banks of the Little Harpeth River in mid-Tennessee? (No wonder these ugly critters have grown so apathetic they're biddin' for a spot on the Endangered Species Chart.)

Didja know that the teensy,weensy Tufted Titmouse has a brain as big as an 18-foot-tall Brontosaurus? But despite the bonding potential of intellectual compatibility, no records exist that suggest the two ever dated or even messed around.

Didja ever wonder what the purple billed sap sucker eats after the raspberry season?

This here is Vernon's Laughing Gull—a rare seashore breed sighted only once by my ol' buddy Vern after he polished off his second six-pack after a long, hot day on the Carolina beach.

CAN YOU IDENTIFY THIS DISTINCTIVE BIRD TRACK?

It's the unusual mark of a Rhode Island Tern which has jammed its right foot in a martini glass left on the beach by a careless picnicker.

This frisky, friendly Surf Sandpiper —only 7 inches tall— emptied the beaches during a bizarre panic on Long Island in the Summer of '84.

The famous Corkscrew Coot drank nothing but estate-bottled wine but it suffered desperately during the 1970s when metallic caps became popular on jug wines.

(See Endangered Species Chart.)

THINGS I LIE AWAKE WORRYING ABOUT

1. What Mt. Rushmore would look like if they added Fred Flintstone's face.
2. Who measured the equator and did he or she get hot.
3. Where the Sandman gets all his sand—is he buying it at the hardware store or stealing it from my favorite beach? And where does he go during the day?
4. If the silver lining is on the top or the bottom of the cloud.
5. If Santa Claus lets the elves use power tools without adult supervision.
6. What Cher is going to wear next.
7. What we could possibly build with cigarette butts.
8. What to call the water between California and the rest of us after the earthquake—should it be the Baja Channel or the Gulf of Nevada?
9. How often Superman washes his Super outfit.
10. If George Washington and Abraham Lincoln mind sharing Presidents' Day.
11. If those soft drink pull tabs that you push into the can are dirty.
12. If Barney Fife is happy living in Mt. Pilot.

37 THINGS I HAVE WONDERED

1. How do we know the sun is 93 million miles from here? How do you measure a thing such as that?

2. Is it always darkest before the dawn?

3. Do the good die young? And should I worry?

4. What do they really put in hot dogs? And should I worry?

5. What weighs more, a pound of steel or a pound of dream whip?

6. Did the man who thought of putting the candy and gum next to the grocery check-out line have any children of his own?

7. If all the snow and ice melted in Alaska, would it still be the largest state?

8. Do wet ducks fly at night?

9. Why dogs have that peculiar way of greeting each other.

10. Why is there an interstate in Hawaii?

11. Where does Dow Jones live?

12. What other things has the Surgeon General determined?

13. Does Jack Frost have a zip code?

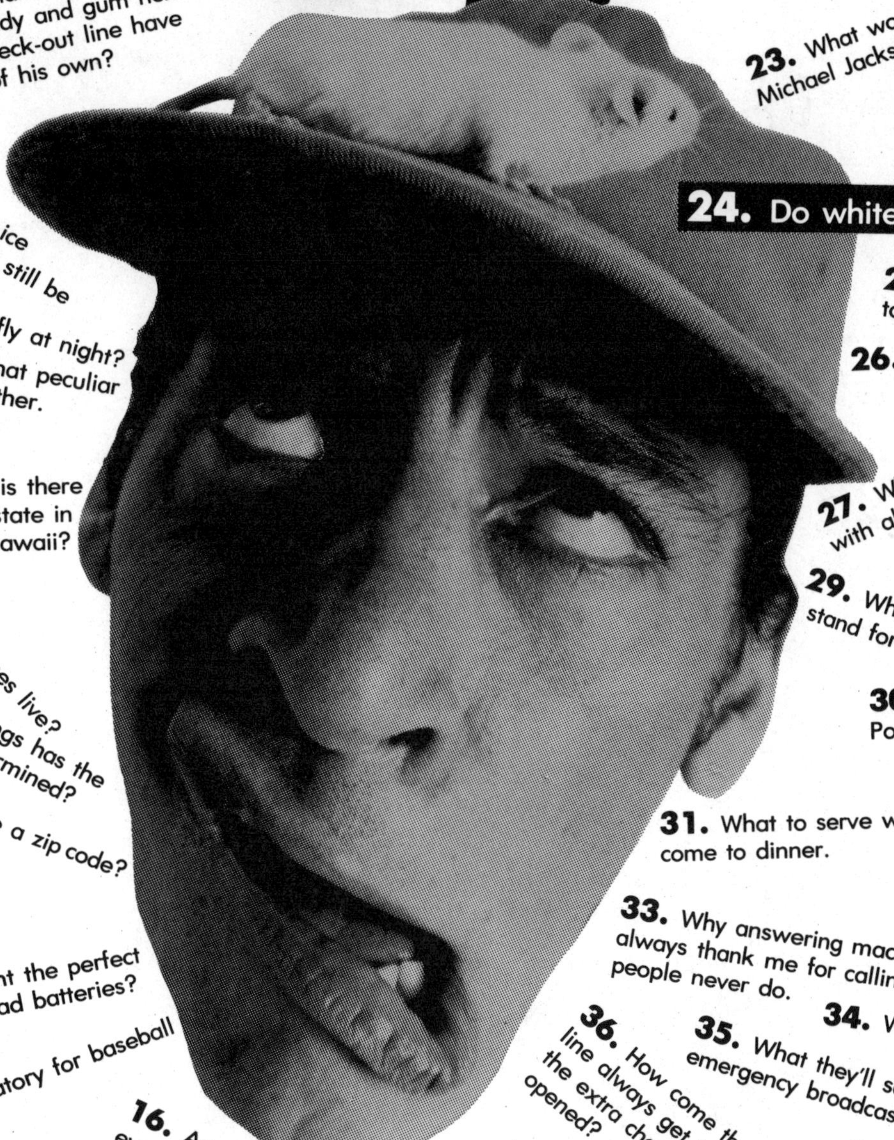

14. Is a flashlight the perfect place to store dead batteries?

15. Is it mandatory for baseball players to spit?

16. Are carrots good for your eye sight?

17. What would life be without the 3-prong adaptor?

18. Why they call New Jersey the Garden State.

19. What will happen if you remove the little tab from the bottom of a new mattress?

20. Are fat people jolly?

21. Do elephants never forget?

22. Who was the first Pro football player to say "Hi, Mom" on national TV?

23. What would it be like to be Michael Jackson for a week?

24. Do white mice cause cancer?

25. Does ugliness go clear to the bone?

26. Are bulls color blind?

27. What does the tooth fairy do with all those teeth?

28. Why do they call it Greenland?

29. What does R.S.V.P. stand for?

30. If there would ever be a Pope LeRoy.

31. What to serve when "dorks" come to dinner.

32. Who are dorks?

33. Why answering machines always thank me for calling but people never do.

34. What Rotarians do.

35. What they'll say when the emergency broadcast isn't a test.

36. How come the last people in line always get to be first when the extra check-out counter is opened?

37. How far is it from New York to the sliding board?

? ? ? QUESTIONS I NEED TO ASK SOMEBODY ?

- Whatever happened to Mr. Roebuck?
- How many used razor blades does that little slot in the back of your medicine cabinet hold and if you buy an older house is it already full and will you get arrested if you drop a new razor blade inside?
- Is L. L. Bean a boy or a girl and what does the L. L. stand for? Lucky Lad? Lucky Lass?
- Where is the Shriners' shrine and what does it look like?
- Will the grass still be greener on the other side of the fence if you use fertilizer on your side?
- Will you like your English muffins, French toast, and Danish pastry more if you talk with an accent?
- If women can join the Army, the Navy and the Marines, can a man be a W.A.C.?
- Who figured out that everybody really has blue blood until the air hits it and then it turns red and why does the air want to hit it and why didn't they pick another color so you wouldn't get sick seeing it?
- Do cloth tape measures really stretch or is this a story made up by the people who make metal tape measures?
- Can anybody really eat just one piece of popcorn and do they just not like the movie?

Tattoo this on your brain, Vernon: Crazy L, Postage Stamp, Four Corners, "T," Diagonal, Picture Frame, Black Out.

We're talking Bingo, buddy, B—I—N—G—O!

Bingo just happens to be one of the fastest growing sports in America, not to mention a great way to supplimentate one's income while having more fun than the law allows.

This game of skill has the longest season of any major sport. It's played every night of the week, all round the country, in schools, churches, Masonic Halls, Moose Clubs and wherever Toast Mistresses meet.

I know by now, Vernon, you're wondering who the heck plays on all these Bingo teams, and how the heck Bingo competes with prime time TV and the big leagues like the A-Team. Well, Vern, it's folks like you and me that make Bingo what it is today—lay people going for the gusto, bowlers, wrestling fans, stock car racers, K-Mart shoppers, and nearly everybody but Indianapolis Colt fans.

Bingo is inspiring an upstanding group of armchair athletes . . . Americans seeking the Dream . . . training for the pay-offs . . . their eyes on the Gold . . . and we're not talking Olympics, KnoWhutImean?

BEWARE OF RECYCLED TRENDS

WAYS TO SCARE FISH

INDOORS

1. Tap on the glass. Yes, fish are extremely nervous and this only makes things worse.
2. Don't feed them for 3 or 4 days. Then dump a lot of food in the water at one time. This increases their dependency on you and makes them worry about having anything to swallow with their water.
3. Screw up your face and try to look like a kissing fish. The real kissers will be terrified of your humongous lips and the other fish will back into parked cars.

OUTDOORS

1. Tap on the glass.
2. Feed them pork rinds, coleslaw, and bread crumbs without the edges cut off. In minutes they will develop indigestion and snap at anything else you send down, even if it's on a hook.
3. Talk. To fish, English is a foreign language and we all know how scary it is to hear someone using words we don't understand...comprendo deja vu?
4. Drop the anchor as fast as you can. You could hit one and have him for dinner. Of course, this will be the only fish you have for dinner because the others saw it happen and will try you for murder.
5. Rock the boat. There is nothing more terrifying than being trapped underwater with a huge black thing overhead swaying back and forth. What's even better is if you can manage to fall out. Fish are absolutely neurotic when anything else gets as wet as they are; they're afraid they'll have to trade places and get dry.
6. Show them Alfred Hitchcock movies. "The Birds" oughta do the trick.

HOW TO

How To Predict Weather From Animal Behavior

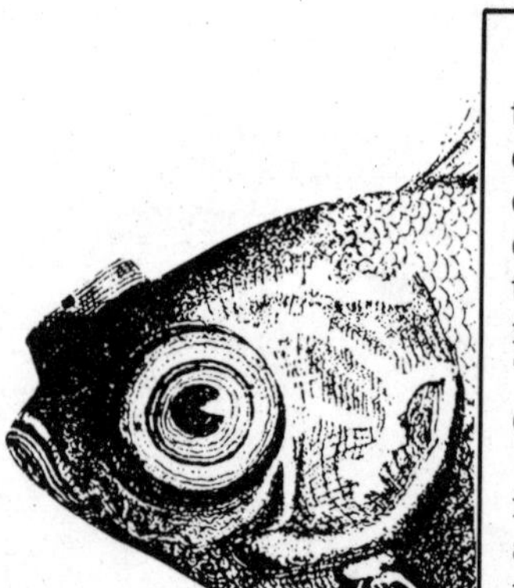

We can't all live in the country and watch the wooly caterpillar so here are things you can do at home. You know that it will be a cold winter if you leave your gold fish out overnight and in the morning when you touch your tongue to it, it sticks. That's bad news for your gold fish and your tongue. The good news here is that it isn't picture day at school.

Another good way to predict a cold winter is to put your Mexican hairless dog outside, and if he runs around with his tail tucked under his belly pointing South, you know its gonna be a cold winter. If after a minute or so he hasn't run around at all, it's probably winter already and your dog has frozen to death. If you don't have a Mexican hairless, you could shave your terrier to perform the same experiment. Before you do that, send for Ernest P. Worrell's Home Remedies and read how to treat small animal bites, KnoWhutImean?

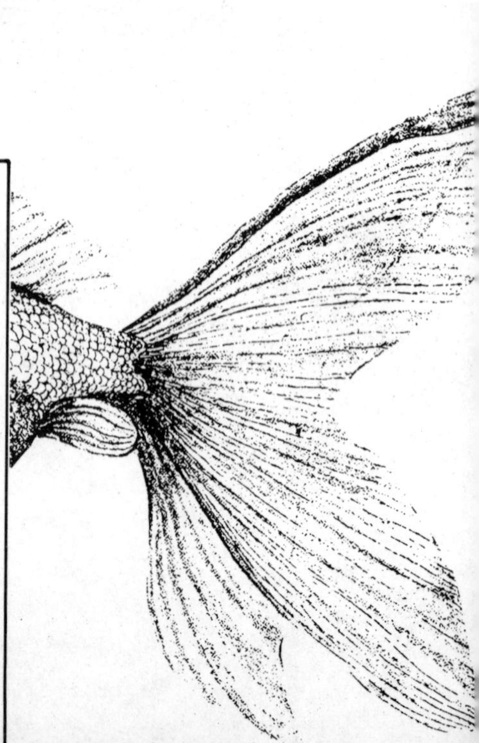

HOW TO BREED RABBITS

Betcha didn't think I knew how to breed anything but power tools with small electrical appliances, did ya, Vern? But your old buddy Ernest is vers-a-tile, KnoWhutImean?

First, you fix up a big bowl of tossed salad and set it on the porch. It's best not to use onions, but do saturate it in ranch dressing. Then wait for the rabbits to drop by. If they don't, you probably live in the city so you'll have to use a congealed salad. While the rabbits are eating, pick out the ones that look well-bred and put them on a high shelf. You can tell a well-bred rabbit because it doesn't chew with its mouth open. Keep this up 'til you have as many rabbits as will fit on the shelf.

Once you've raised the rabbits, you're in business, old buddy. You can cook them and serve Rabbit On A Stick or Rabbit Nuggets. You can use the skins to go into the profitable hair-weaving business in your own home. Of course, rabbits make wonderful pets in case you are stranded on a desert island with only your pet and no food. And there's always a big consumer demand for lucky rabbits feet on a key chain. Why they believe they're lucky I'll never understand because they weren't very lucky for the rabbit and he had four of them. Maybe it's got something to do with the key chain, KnoWhutImean?

How To Pass A Physical

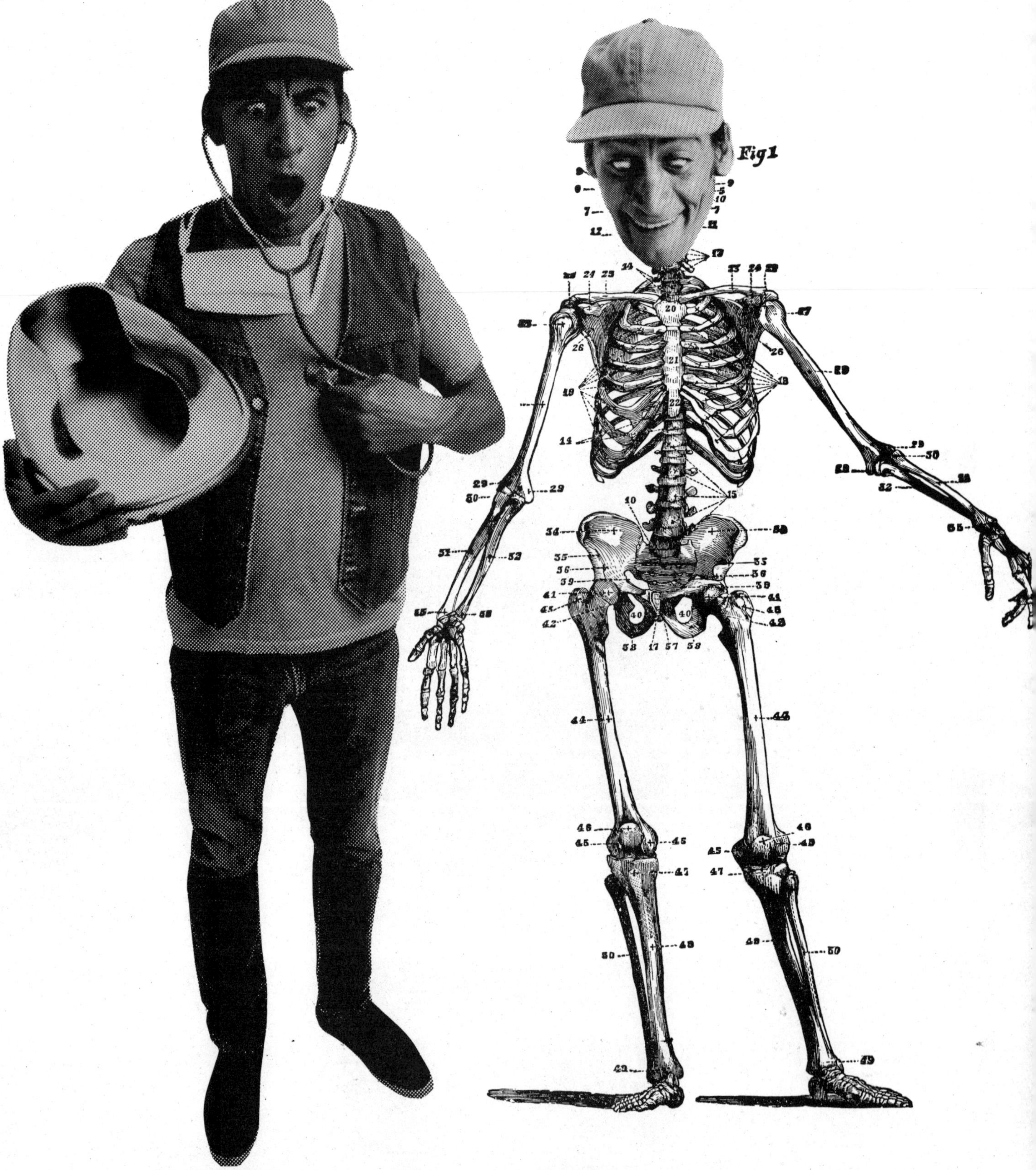

1. Wear new clothes—it makes a healthy impression.
2. Show an interest—ask the nurse questions like "How 'bout that traffic this morning" or "Do you take sugar with your tea?"
3. Wear clean socks—enough said.
4. Breathe in when you stick out your tongue.
5. Don't pretend to be taller than you are.
6. And Vern, I wouldn't mention that uncle of yours that had a cabbage-like thing growing out of his neck, KnoWhutImean?

SNAKE BITES

WHAT TO DO.

Here's what to do as soon as you notice that a snake bit you:

1. Tie a rag above the bite.
2. Cut an X over each bite.
3. Suck out the poison.
4. Go to a hospital and tell them you drank poison.

SUB NOTE: Biting a snake is not recommended. It usually makes them mad and first aid is tricky. However, when a bite does occur, apply rags as follows:

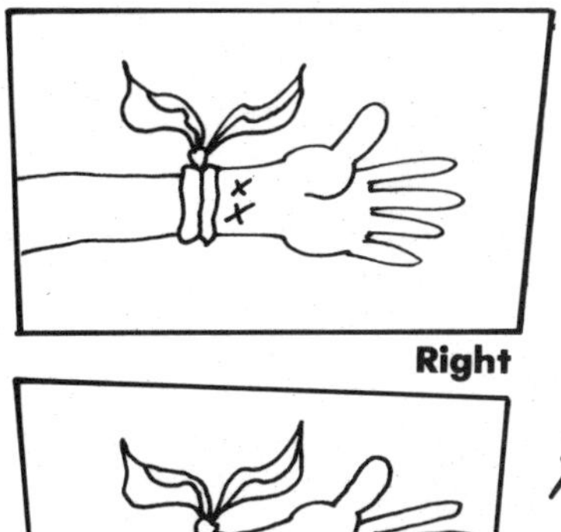

Right

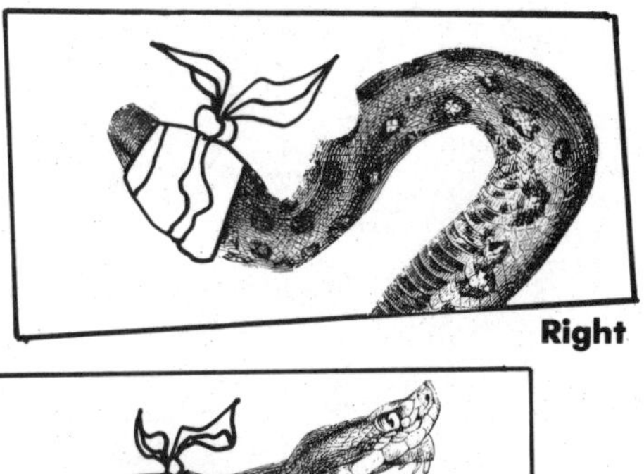

Right

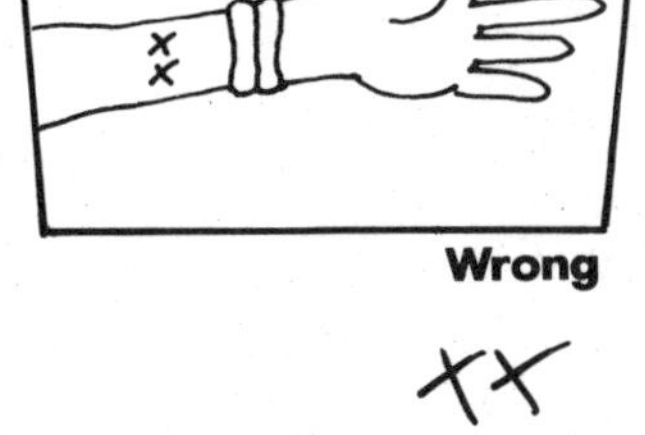

Wrong

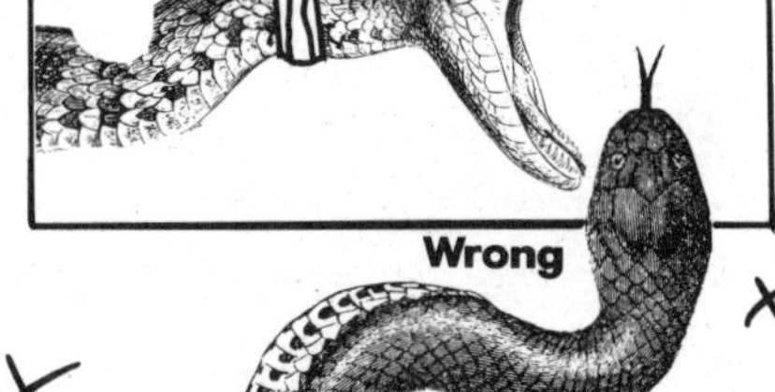

Wrong

How To Use Refrigerator Magnets

First off, remove your entire decorous collection from the box. When all those scraps of paper fall, shove 'em underneath, especially the reminders about going to the dentist and paying the phone bill.

Then line up your magnetized objects d'art and give pause to this thought: why limit your flair for display to the refrigerator? Why not put 'em on the Vegamatic? Arrange a few on the bacon drippin's can. I do not advise embellishing the ice cube trays. Vern indulged in that nouveau idea and got himself into a socially embarrassing situation. After gulping down a cold drink on ice from the above mentioned tray, his dental work was embellished with an exact replica of a broccoli spear. A decorator's nightmare, KnoWhutImean?

MODERN TORTURE DEVICES

Did ja ever wonder who thought up all this stuff??

Speed Breaker Bumps: They always put these teeth-rattling nuisances in places where you couldn't speed anyway.

Frozen Orange Juice: Just try and budge it out of the can without it attacking you.

Revolving Doors: Don't get in the same slot as your buddy.

Non-Dairy Creamer: Stir real fast or you'll get the non-dairy lumps in your coffee.

Car Burglar Alarms: The only people who know how to turn 'em off is car thieves.

Answering Machines: A cruel hoax and a ruse to make you talk to yourself.

Bubble Packages: They seal up a piece of cardboard with a piece of molded plastic and what you want is locked inside and you can't ever get it out.

Tube Socks: Do tubes wear socks? No. And feet don't wear tubes either.

The Miniature Spare Tire: Brought to you by the same folks who invented miniature warranties.

Heat Sensitive Elevator Buttons: In the winter, you're forced to use your tongue.

Creative Financing: It's amazing how many clever things they can think of to do with my money.

Name Tags: What a great party! A room full of people staring at each other's lapels.

Golf on TV: Who can handle that kind of tension?

Pedestrian Crossing Buttons: Sure, they're gonna let someone without a car control the traffic light.

Multi-Screen Movie Theaters: These might be okay if they included multi-snack-bars.

Dehydrated Beef Stroganoff: You have to sit around the campfire eating reconstituted hash while your buddies barbecue steaks.

Computerized Restaurant Checks: Who ordered the BLI? And did I get two TBA's?

Machines that make change: If you forgot to iron your $1 bill, you can forget the change.

IT'S THE ERNEST P. WORRELL INTERNATIONAL

FAN CLUB

THE REAL DEAL

Golly-
Bob-
Howdy

Big Fun.

Be the first one in your immediate area to become a gen-u-ine, official, bona fide member of the "Love Ernest" Team!

For a measly $5.00 (cheaper than the Culture Club), you get:

★ your very own official membership card
★ an autographed picture of me, your ol' buddy Ernest, in living black-and-white
★ the "KnoWhutImean News"—a newsletter that's full of articles, photographs and fun facts about everybody's favorite next-door neighbor, Ernest P. Worrell
★ a two-color, limited edition, machine-printed authorized Ernest bumper sticker
★ information on how to order Ernest posters, T-shirts and other life-giving souvenirs

Sign up today, Vern! 'Cause you could use some fun in your life...KnoWhutImean?

And it only costs $5.00. That's five $1 bills, or twenty quarters, or 500 pennies but whatever you do, *DO NOT SEND CASH!*

→ *DO* send your real name and address and your check or money order for $5.00 to:

Ernest Fan Club
P.O. Box 23325 • Nashville, TN 37202

Hey, Vern! Just what you've always wanted...Me!

Full Color
POSTER
Humongous 20″×29″ color portrait of your ol' buddy Ernest, complete with smaller (but just as cute) poses at the bottom. Rolled and delivered in a mailing tube.
$5.00 EACH

KnoWhutImean?
BUMPER STICKER
We're talkin' two wonderful colors, all on a sticker 10¾″ long. 1 size fits all cars. $2.00 EACH

Official Ernest
2″ BUTTON
The same two wonderful colors with—who else—ME in the middle. A must if you like circles! $1.50 EACH

Life-Size Color
MASK
Who could resist a face like this? Full color, full size, and full of laughs. An un-average face to wear in your average neighborhood. 12″ high, bagged with a genuine elastic string.
$2.00 EACH

KnoWhutImean?
Color **T-SHIRT**
Cotton, polyester, 'n me—all on a clean white T-shirt with red neck and sleeve trim. Full color and ready to wear. Be sure to specify your size with your order, so your ol' buddy Ernest won't have to take a wild guess! KnowWhutImean?
$8.00 EACH
Adult Size
S, M, L or XL

"HEY, VERN! IT'S MY FAMILY ALBUM" TV COMEDY SPECIAL **PLUS** ERNEST 'N VERN'S GREATEST TV COMMERCIALS
½″ VHS VIDEO CASSETTE
We're talkin' big fun now! Turn me on any time you like with this long playing home video cassette! First on the tape is my own TV Comedy Special—approximately 49 minutes of madcap comedy—starring me as me, plus me as several of my infamous relatives!
Plus you get the very, very best Ernest 'n Vern commercials in the world, each 30 seconds in length. Yeah, Vern, this tape's got more laughs than the 4th grade's got spitballs!
½″ VHS Cassette.
$30.00 EACH

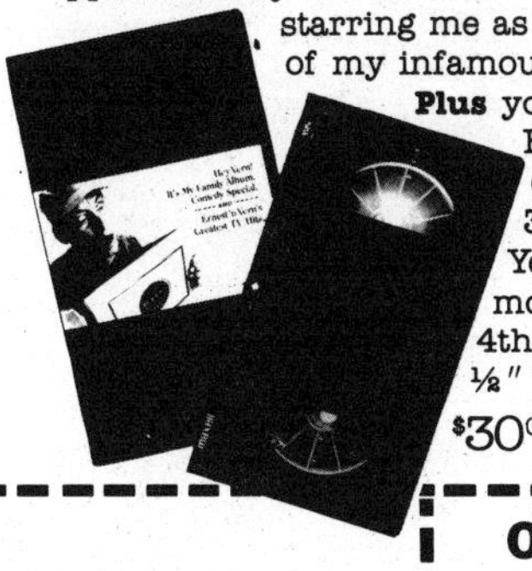

Peel-Off Ernest
STICKERS
A 5″×10″ sheet of 10 crazy faces, all in color, all made by me, and all printed on your popular peel-off sticker paper. Wrapped in a disposable plastic bag!
$2.00 PER SHEET

ORDER FORM

Please print as if your order depended on it...
All prices include shipping and handling within continental U.S. Allow 2-3 weeks for delivery.

Name ____________________
Address ____________________
City ______ State ______ Zip ______

Qty.	Description (Size of T-Shirt)	Price Each	Total Price

Subtotal $ ______
Tennessee addresses add 7 ¾% sales tax $ ______
TOTAL $ ______

Make checks or money orders payable to:
Ernest Fan Club • P.O. Box 23325 • Nashville, TN 37202

ORDER FORM

Please print as if your order depended on it...
All prices include shipping and handling within continental U.S. Allow 2-3 weeks for delivery.

Name ____________________
Address ____________________
City ______ State ______ Zip ______

Qty.	Description (Size of T-Shirt)	Price Each	Total Price

Subtotal $ ______
Tennessee addresses add 7 ¾% sales tax $ ______
TOTAL $ ______

Make checks or money orders payable to:
Ernest Fan Club • P.O. Box 23325 • Nashville, TN 37202

ORDER FORM

Please print as if your order depended on it...
All prices include shipping and handling within continental U.S. Allow 2-3 weeks for delivery.

Name ____________________
Address ____________________
City ______ State ______ Zip ______

Qty.	Description (Size of T-Shirt)	Price Each	Total Price

Subtotal $ ______
Tennessee addresses add 7 ¾% sales tax $ ______
TOTAL $ ______

Make checks or money orders payable to:
Ernest Fan Club • P.O. Box 23325 • Nashville, TN 37202

STUPID QUESTIONS

- Am I dressed all right?

- Does that dog bite?
- Is it going to rain?
- Is it bigger than a breadbox?

- I'd like a dollar's worth of stamps, please. How much will that cost?
- Do you want this done right or should I come back later?

- Have you been sick or did you get a hair cut?
- Is that what you're going to wear?
- What's my home phone number?

- Does it look like I have stupid written all over my face?
- Do you want the rest of your french fries?
- Where did ya lose it?
- How long has it been?
- Have you ever seen one of those things up close?
- Where's your timesheet?
- Can I treat?
- Where's the red deck chair?
- You want it when?
- Are you asleep?
- Is anybody sitting here?
- Are you awake?
- Is it soup yet?
- Are we there yet?
- Is it free?
- That's a big elephant, isn't it?

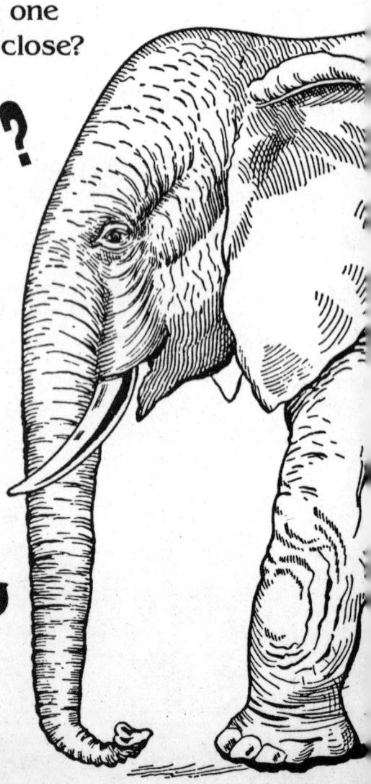

- KnoWhutImean?

THINGS YOU CAN DO WITH CAULKING
1. Create flexible maps of the world for your science project. Paint part of the caulking blue and you have an ocean.
2. Put a little bit on every edge of a pinecone. You won't end up with anything but a pinecone with caulking on it, but it will pass the time.
3. Make up interesting desserts. Squirt it on top of strawberry shortcake or spoon it into tall glass parfait dishes and pretend it's vanilla pudding. Be sure to serve it before it hardens or people will guess.
4. Seal the bedroom door shut while your sister is inside playing with her Barbie dolls.
5. Roll it into little balls in the palm of your hand, let them dry, and use them to replace the marbles that are missing from your Aggravation game.
6. Decorate your Christmas tree. It's the same color as popcorn but it won't attract mice.
WHAT TO DO WHEN YOU RUN OUT OF CAULKING
Use silly putty, vanilla pudding or cottage cheese—no one will ever know the difference.

ARE YOU NORMAL¿

(A Pop Quiz)

1. When you stand on your head, the change falls out of your pocket.
   ☐ True ☐ False
2. You have toyed with the idea of changing your name.
   ☐ True ☐ False
3. You were born too soon.
   ☐ True ☐ False
4. You feel real dumb when you dial the wrong number.
   ☐ True ☐ False
5. There's an ice tray in your freezer with only one cube in it.
   ☐ True ☐ False
6. You do not know the capital of North Dakota.
   ☐ True ☐ False
7. You left the bathroom light on.
   ☐ True ☐ False
8. When the directions say "you can't miss it", you always do.
   ☐ True ☐ False
9. You haven't looked at your elbows in days.
   ☐ True ☐ False
10. Even when nobody's looking, you try not to look dumb.
    ☐ True ☐ False
11. Your favorite place on elevators is next to the buttons.
    ☐ True ☐ False
12. You secretly envy people who look good in fuchsia.
    ☐ True ☐ False

YOUR SCORE: Add all the True answers, multiply by the False answers, and divide by your shoe size. If the number is greater than 47, you are probably good at math.

You Can Cheat On This Part

1. You balance your check book
   ☐ almost sometimes
   ☐ slightly less than sometimes
2. Your brothers and/or sisters are
   ☐ taller than you
   ☐ shorter than you
   ☐ all and/or none of the above
3. You squeeze the toothpaste
   ☐ from the middle
   ☐ from a distance
   ☐ from memory
4. Which best describes the unmatched socks in your drawer
   ☐ 10
   ☐ too short
   ☐ argyle
5. Your favorite thing on TV is
   ☐ The 3 Stooges
   ☐ The horizontal control
6. You think you're so smart because
   ☐ you buy dry dog food in 100 pound sacks
   ☐ when you take the last cup of coffee, you never make another pot
   ☐ you put peanuts in your cola bottle

YOUR SCORE: If you needed to have more choices to these questions, you are a multiple personality.

Add 25 words to the following and write small:
I absolutely had to park in the handicap spot because________________.
You score this one on the honor system.

THINGS PEOPLE ONCE THOUGHT (BUT NOW KNOW BETTER)

1. All shoes are really made by elves on the night of the full moon.
2. You should feed a cold and starve a fever.
3. Icebergs are melting faster than they should.
4. There is no name for the little thing that looks like a punching bag and hangs down in the back of your throat.

5. Chickens never sleep.

6. When you sneeze, your heart stops.
7. Hot water freezes faster than cold water.
8. If you borrow money on the 1st day of May and don't pay it back by the 1st of June, your hair will fall out.
9. Dogs don't sweat.

10. Cornbread should never be cooked in a round pan.

11. That dogs slept on their backs in winter and on their stomachs in summer. Now we know this is true only when the stellar configuration according to your vernal equinox is in alignment with Pisces on the half shell. Or vice versa, I forget.
12. That cats would sneak into the baby's room and suck out her breath when she was asleep. That's not true and never was. But it's still a good idea not to trust a cat too far... any cat.

SOME
OTHER
PAGE

HOW TO TALK LIKE ME

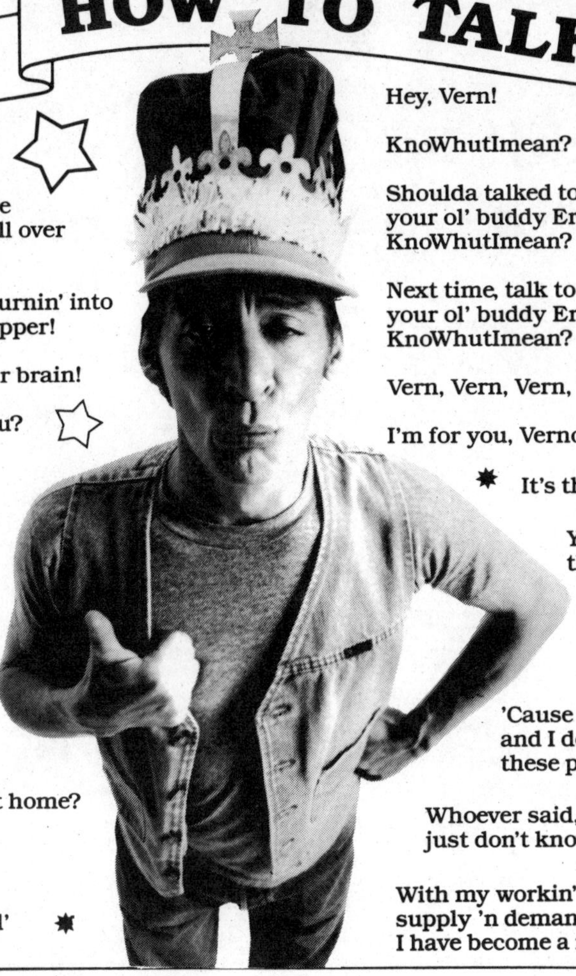

Be there
or be square!

Do I look like I have
"stupid" written all over
my face?

Looks like you're turnin' into
quite the little shopper!

Tattoo that on your brain!

Comprendo deja vu?

Golly-bob-howdy!

It's a lean, keen
news machine...
KnoWhutImean?

An informed fan
is a happy fan...
KnoWhutImean?

Prime time
all the time!

Why roam
when the best is at home?

Local news
is where it's at!

Look what mean ol'
Mr. Gravity did!

Hey, Vern!

KnoWhutImean?

Shoulda talked to
your ol' buddy Ernest first...
KnoWhutImean?

Next time, talk to
your ol' buddy Ernest first...
KnoWhutImean?

Vern, Vern, Vern, Vern, VERN!

I'm for you, Vernon!

It's the real deal...KnoWhutImean?

Your taste buds'll be so happy
they'll be asking for mink coats!

Your ol' buddy Ernest notices these
kinds o' things, Vernon...
Must be my F.B.I. trainin'...KnoWhutImean?

'Cause if you don't pitch in your part,
and I don't pitch in my part,
these poor people won't have a part to pitch in!

Whoever said, "money can't buy happiness"
just don't know where to shop...KnoWhutImean?

With my workin' knowledge o' cash flow,
supply 'n demand, 'n the free enterprise system,
I have become a noted entre-manure...KnoWhutImean?

Whatever it takes,
wherever it breaks!

If you're waitin' on me,
you're backin' up!

Y? Why not?
KnoWhutImean?

Hey, Vern!
Wear me out!
KnoWhutImean?

I got a nose for news...
KnoWhutImean?

Movin', movin', movin'...
Keep those viewers viewin'!

Why, it's just like
Mom used to make!

The bus
is a must
for guys
like us!

ERNEST'S PERSONAL FAVORITE ALL-TIME PHRASEBOOK

Efficiency don't mean nothin' to a pig!

Financin' don't mean nothin' to a pig!

Style don't mean nothin' to a pig!

Appliances don't mean nothin' to a pig!

Tax shelters don't mean nothin' to a pig!

Time don't mean nothin' to a pig!

Trust me on this one!

Save mucho dinero...comprehendo?

Put this 'un here
'n that 'un there
,n that 'un over there!

Heat pump, shmeat pump!

If they ain't got it,
you don't need it!

It'll make you a lean machine...
KnoWhutImean?

Looks like it's got a shoo-oo-oort in it!

Oh, *there's* your problem, Vern!

My daddy useta work on them.

I'm a clean machine, jellybean!

Zappo!

Without condition...
el unconditiono guaranteedo.

Be good to your ears
'cause they outnumber us 2 to 1!

Any closer to heaven
'n you'd have to wear wings!

Primo!

Back 'em, dock 'em, spot 'em,
move 'em down the highway!
We're talkin' White Line Fever!

Mornin', Glory!

Do yourself a flavor!

Whatcha havin' for breakfast?

Whatcha havin' for lunch?

Whatcha havin' for dinner?

Whatcha havin' for dessert?

It's absolutely perfect for
us "health nuts"...KnoWhutImean?

Play it smart, Vernon.

You right as rain, buddy!
How'd you get in here?

What'll they think of next?

Hot Stuff...KnoWhutImean?

Nice muumuu, Mrs. Johnson!

It's got that little
splash o' color...
KnoWhutImean?

Any closer to Easy Street
'n you'd have to change Zip Codes!

We're talkin' skinny city!

We're talkin' bidness!

We're talkin' Pizza Logic here.

We're talkin' show bidness!

We're talkin' ver-i-table extravaganza!

We're talkin' el less-o dinero, comprendo?

We're talkin' cash flow...
comprendo?

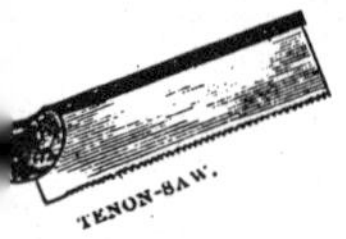

Do-It-Yourself Guide For Installing A Lifestyle In Your Home

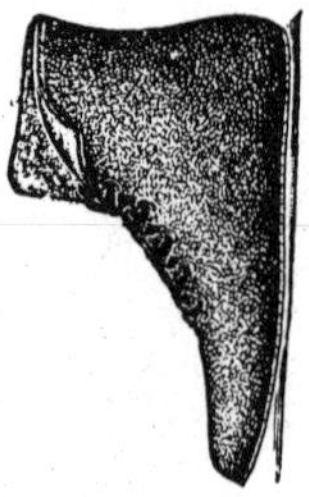

Why waste your money on those pre-assembled Lifestyles available in your finer shopping malls? They fall apart anyway. Yours truly has constructed a Lifestyle using materials laying around the house. And you can't tell it from the genuine article. Besides which, I made it so it fits into a seldom used corner of the basement. That way, I don't have to fool with it much except when somebody else with a Lifestyle comes to visit. That's a primo advantage, because a Lifestyle can eat you out of house and home if you let it take over. Now, when I want to keep up appearances as a Man-a-la-Mode, I just go get my portable Lifestyle out of the basement. (Never install a permanent Lifestyle. Next thing you know, it won't be a la Mode anymore.

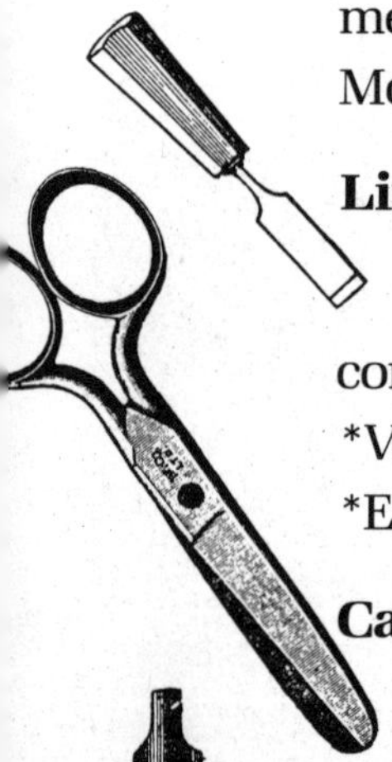

Lifestyle Construction Materials

*Old newspapers and worn-out socks for the filling—it should be soft and comfy.
*Velcro, so you can get in and out of it fast.
*Ecru paint, so it won't clash with the linoleum.

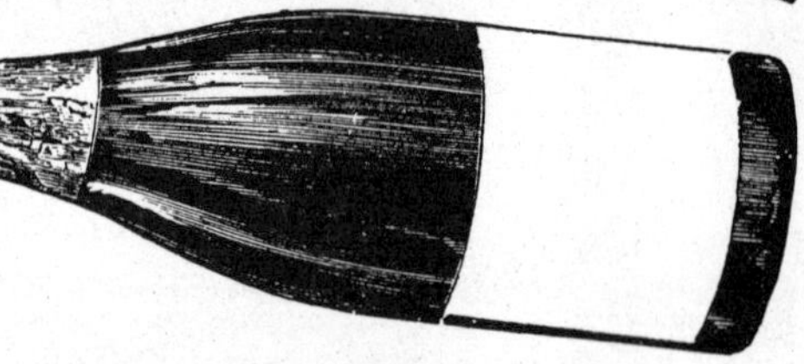

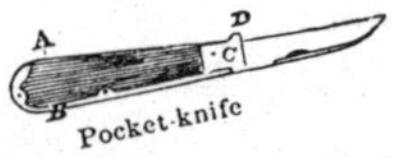

Care & Feeding of Your Lifestyle

Most Lifestyles can survive on recycled aluminum or anything that shines a lot. If your Lifestyle says it can only drink water bottled in France, give it a bicarbonate in a Coke bottle—it can't tell the difference.

Displaying Your Lifestyle

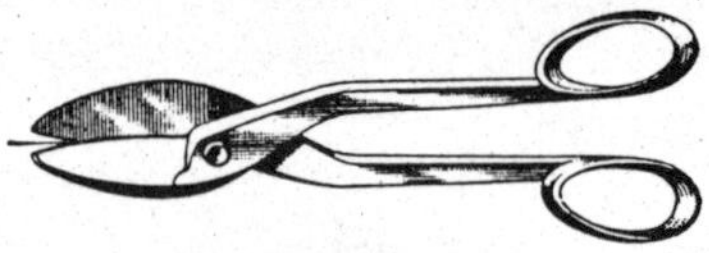

*In hot weather, keep your Lifestyle in the back yard and rig the garden hose to it. Then refuse to let the neighbor's kids swim in your pool.
*For pot luck dinners, make your Lifestyle wear an apron that says, I'd rather be dining a la carte.
*For short trips, you can save space by wearing your Lifestyle — with your best pair of sneakers.
*For long trips, make your Lifestyle carry the luggage.

Everyday Lifestyle Usage

*To get rid of a salesman, tell him you can't buy his gizmo because it doesn't fit your Lifestyle.
*On tax forms, claim an extra deduction for supporting a Lifestyle.
*When you want to loaf around, hide under your Lifestyle.
*If the neighbors complain, blame it on your Lifestyle.
*If your Lifestyle complains, change it for a new improved one.

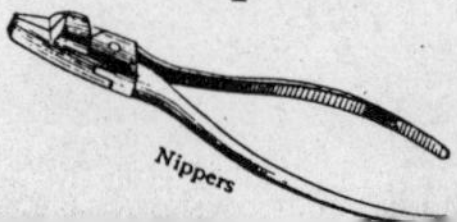

People To Avoid

We're known by the company we keep, Vern. Birds of a feather flock together

People who never have change. What I have spent on parking meters and tips alone could feed a family of six for a year.

People who are accident prone. When they slip on the ice, they'll grab on to you and you'll fall down with them.

People who don't return things they borrow. I'll get that lawn mower back to you as soon as I get it running again.

People who don't remember your birthday. Mine was last week, by the way, Vern. I guess you must have mail-ordered me something that hasn't got here yet.

People who throw bottles at mirrors behind the bar. Seven years of bad luck is a lot to withstand.

YOU A-MAZE ME!
PART THREE

FLOOR PLAN OF VERN'S HOUSE

See if you can help Ernest get from the front door to the back yard without bumping into any walls. If you can, you're doing better than ole Vern himself!

MORE PONDER... THINGS TO ASK YOURSELF EVERY MORNING

Has the dog knocked over the trash can?

If things were different, would I notice?

Did the mail come yet?

Should I flush twice?

Will it rain on my vacation?

Do these socks match?

Will the check bounce?

Should I take an umbrella?

Is 4 enough, is 7 too many?

Where is my other shoe?

Why does popcorn always taste better at the movies?

Where the Sam Hill is Sam Hill?

Is the milk sour?

Did I leave the water running?

Which one is the cordial cherry?

Will Vern remember my birthday?

Did I vote for the right person?

How many times have I sneezed in my lifetime?

Are pigs smarter than horses?

Which one is the wrong side of the bed?

Who ate the first coconut ever eaten?

Will the car start?

What they really mean when they say...

1. **May I help you?**
   (Are you going to spend any money?)
2. **Everyday low price.**
   (This doesn't go on sale 'til next week.)
3. **Void where prohibited.**
   (Don't move to Kansas.)
4. **Slightly irregular.**
   (You buy it, you fix it.)
5. **Reliable source.**
   (The person in charge, but he won't admit he said it.)
6. **Glad to meet you.**
   (I didn't catch your name.)
7. **Whatever you say.**
   (A dumb idea like that isn't worth arguing with.)
8. **Hope you're feeling better.**
   (You don't look it.)
9. **You're looking so much better.**
   (You're looking a little less awful.)
10. **Gotta minute?**
    (My battery needs jumping.)
11. **Will there be anything else?**
    (Here's the tab.)

All-Time KnoWhutImean Puzzle Machine!

Check this out, buddy!

There are words within words . . . **comprendo deja vu, thunderbird?**

For example, take my all-time personal favorite, **KnoWhutImean!** Now see how many other words (of at least two letters, ol' buddy, ol' pal) that you can make from the letters in KnoWhutImean, Jellybean!

Ready? Aim? Fire!

(Some Answers on Some Other Page)

Okay, now that you're so durn smart, here's a simple little question that may just stump you silly.

ARE YOU RIGHT-EYED OR LEFT-EYED?

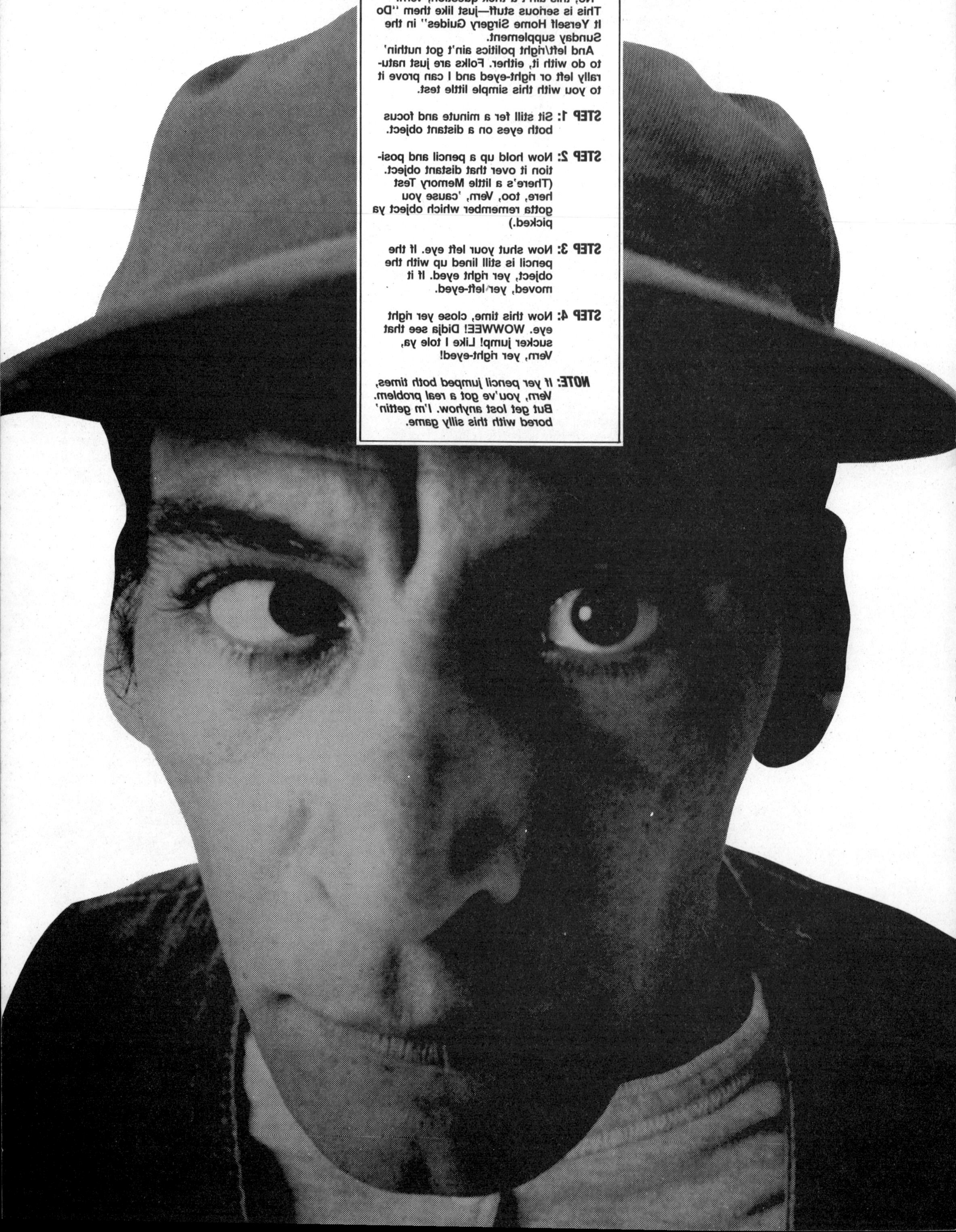

No, this ain't a trick question, Vern. This is serious stuff—just like them "Do It Yerself Home Sirgery Guides" in the Sunday supplement.

And left/right politics ain't got nuthin' to do with it, either. Folks are just naturally left or right-eyed and I can prove it to you with this simple little test.

STEP 1: Sit still fer a minute and focus both eyes on a distant object.

STEP 2: Now hold up a pencil and position it over that distant object. (There's a little Memory Test here, too, Vern, 'cause you gotta remember which object ya picked.)

STEP 3: Now shut your left eye. If the pencil is still lined up with the object, yer right eyed. If it moved, yer left-eyed.

STEP 4: Now this time, close yer right eye. WOWWEE! Didja see that sucker jump! Like I tole ya, Vern, yer right-eyed!

NOTE: If yer pencil jumped both times, Vern, you've got a real problem. But get lost anyhow. I'm gettin' bored with this silly game.